C000284936

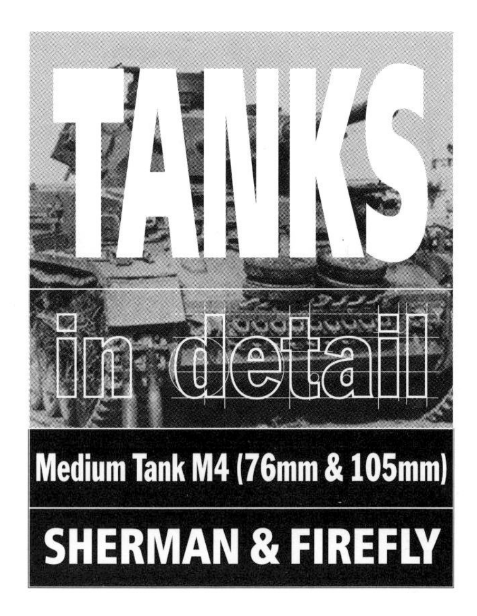
TANKS
in detail
Medium Tank M4 (76mm & 105mm)
SHERMAN & FIREFLY

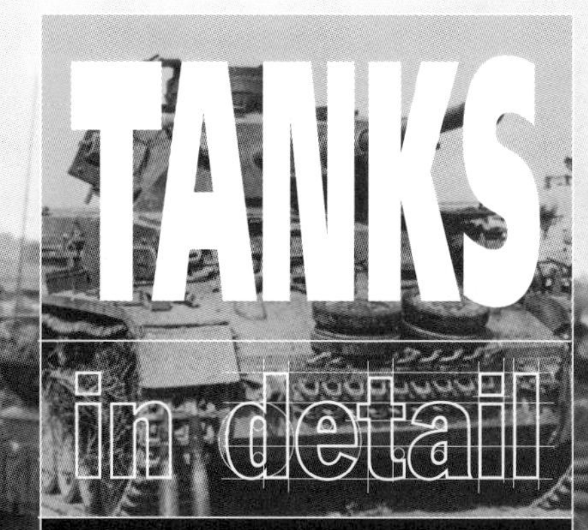

Medium Tank M4 (76mm & 105mm)

SHERMAN & FIREFLY

TERRY J. GANDER

Acknowledgements

This book could not have been produced without
the invaluable assistance of the following people:
David Fletcher, Historian at the Tank Museum *(TM)*,
Bovington, Dorset, England; the Librarian Janice Taite;
the museum's photographer, Roland Groom for original
images and prints from archive material. Also thanks to
John Blackman *(JB)* for original colour images.

Jasper Spencer-Smith
Bournemouth, England
October 2003

Series Created & Edited by Jasper Spencer-Smith.
Design and Illustration: Nigel Pell.
Produced by JSS Publishing Limited,
Bournemouth, Dorset, England.

Title spread: The Canadian Fifth Armored Division on parade at Elde Airport, Holland, on 23 May 1945. Fireflys are prominent with their long 17-pounder gun barrels and also noticeable is the degree of extra protection obtained by appliqué armour and a layer of spare track shoes. *(CA)*

First published 2003

ISBN 0 7110 2989 X

Published by Ian Allan Publishing

an imprint of Ian Allan Publishing Ltd, Hersham,
Surrey KT12 4RG.

Printed by Ian Allan Printing Ltd, Hersham,
Surrey KT12 4RG.

Code: 0311/A3

CONTENTS

CHAPTER ONE

M4 DEVELOPMENT

The Medium Tank M4 development story is complex, due mainly to the number of models involved and the different types of armament installed. The M4 was one of the most important Allied tanks produced during World War Two and it was produced in thousands to fight in every theatre of combat.

The Medium Tank M4, or Sherman, is one of the tank classics. From late 1942 onwards it was always in the vanguard of Allied campaigning and it nearly always seemed to be there when it was wanted. It appeared in quantity, in many forms, variants and sub-variants, so many that to cover them all within one volume is an almost impossible task, especially as updated examples of the M4 series are still likely to be encountered in the 21st century. This account will concentrate on the heavy gun models produced up to 1945.

Development

No account of the development of the heavy gun versions of the Medium Tank M4, known to the British Army as the Sherman, can be complete without a summary of the associated early development history and its production background. The M4 story can be traced back to August 1940 and the initial design sequence that led to the design of the Medium Tank T6, the prototype for the M4 series. Where the T6 differed significantly from its predecessors was that it carried a fully-traversing turret mounting a 75mm gun.

The need for such a vehicle was highlighted by the appearance of the German PzKpfw IV armed with a 75mm gun during the Battle for France in May/June 1940. This debut greatly disrupted the intended US tank production programme for the closest planned equivalent to the PzKpfw IV was then the Medium Tank M2A1, armed with a 37mm main gun. Far-sighted US Ordnance Department planners (and others) appreciated that, despite many strident local political voices to the contrary, the US would sooner or later have to become involved in the European conflict in some capacity or another. When it did it would have to field tanks that were at least on a par with its anticipated opponents. Any future US tank therefore had to be armed with, as a minimum, a 75mm gun.

An interim measure arrived after a period of rapid redesign of the M2A1 and the acceptance that the immediate solution could only carry the required 75mm gun in a limited traverse sponson set into the hull. This was the Medium Tank M3 series known to the British, the main initial users, as the Grant. For full details of the M3 Grant/Lee series refer to Tanks in Detail No 4.

While better than nothing, the M3 series had too many shortcomings for its specified role

Above:
The first of many, the pilot production model of the M4A1 photographed soon after its roll-out from the Pacific Car & Foundry factory on 9 June 1942. The 75mm M3 gun is in the early M34-pattern gun mounting. *(TM)*

which, by 1942, had changed from its original concept of mobile infantry fire support platform to that of battle tank, a war weapon in its own right. This was the armoured vehicle that would have to counter enemy tank activities and serve as the main armoured component during an attack. To its credit the M3 series proved to be remarkable on many counts, not the least being that it enabled the fledgling US tank industry to gear up and gain experience for what was to follow, namely the production of tanks by the thousand. By any measure, the organisation and sudden expansion of US tank production from virtually nothing to churning out hundreds of vehicles every month, growing to thousands at times, formed one of the most remarkable achievements in the history of mass production. It was also a factor which dominated the subsequent Medium Tank M4 story until 1945.

The M4 series prototype, the Medium Tank T6, carried over many mechanical systems and sub-systems from the M3, such as the lower hull, engine installation, suspension and tracks. (The Medium Tank M3 was itself a hasty and drastic revision of the even earlier and short-lived Medium Tank M2A1.) These systems and sub-systems had proved themselves more than adequate on the M3, while whatever major and minor faults that were bound to arise when any new item of military equipment entered combat service had been largely ironed out by the time M4 production began.

In addition, the M3 and the later M4 series benefited greatly by an informal association between the US military authorities and their British counterparts who, by 1941, had already accumulated a useful store of knowledge gained from their experience of armoured vehicle warfare. This led to many modifications suggested by the British being incorporated into the tanks on US production lines.

There was little likelihood that Britain's hard-pressed industrial base could manufacture the tanks required in the necessary quantities. The United States of America could. It was also able to produce the goods within a remarkably short space of time. Following a period when even the most basic tanks somehow managed to make military vehicle designers spend years bringing their progeny to the necessary standard of reliability and combat-worthiness, the prototype Medium Tank T6 was rolled out in September 1941, just six months after the engineering drawings were completed.

Above:
The starting point: Medium Tank M2, armed with up to eight 0.30-calibre M1919A4 machine guns (two could be mounted on the turret sides) and a single 37mm gun. *(TG)*

Right:
A Medium Tank M2A1 giving an indication of its obstacle-crossing capabilities before an audience of US Army officers, many of whom, no doubt, had, never seen a tank before. *(TM)*

Above:
The Medium Tank T20 series was developed as an overall improvement on the Medium Tank M4 series and its possible eventual replacement. This is the T20E3 with torsion bar suspension and a 76mm gun. Development of this series ceased during 1944. *(TM)*

It was the herald for 49,234 tanks. Standardisation as the Medium Tank M4 had been completed by the end of the same month. Production commenced soon after. The first examples went into action at El Alamein in October 1942 as the British Sherman (officially General Sherman).

The M4's El Alamein combat debut was greatly assisted by another international agreement factor, the passing of the Lend-Lease Act by the US House of Congress in March 1941. By this Act the US assumed the burden of supplying (and paying for) the vast quantities of arms, including tanks, that were by then desperately needed by the British and Commonwealth allies. It has to be remembered that in March 1941 the US had yet to enter the war. The wording of the Act was complex but what it boiled down to was a fiction that the US would supply the British with the necessary war material for them to continue hostilities, on the understanding that it would be given back when the war was over. More nations were later added to the Lend-Lease distribution list, including the Soviet Union following Operation 'Barbarossa', Germany's invasion of mid-1941.

For the US tank industry the Lend-Lease Act meant that whatever it was already attempting to produce to equip the US armed forces would have to be numerically doubled. To compound the production problem, Japan attacked the US Pacific Fleet at Pearl Harbor on 7 December 1941, and the World War Two became truly global. The projected US medium tank production target then became over 2,000 tanks for every month of 1942. That total was never achieved but the shortfall was more than compensated for by the diversion of production facilities towards other related vehicles such as self-propelled artillery carriages and other combat vehicles based on M4 chassis.

It is in this context that the 1942 to 1945 Medium Tank M4 programme has to be placed. It also has to be placed in a national production context, for the M4 was developed and manufactured at the same time as the US was undertaking many other gigantic manufacturing programmes. These covered the entire military spectrum, from naval and Liberty Ship

Above:
The Medium Tank T23 was developed in parallel with the T20 series and featured a 76mm gun. Although considered as too heavy, the main contribution of the T23 to the M4 series was that its final turret, armed with a 76mm gun, was carried over to the late-production M4. *(TM)*

Right:
The Medium Tank T23 was considered for limited production but it was never standardised as it offered few advantages over the M4 and had too many troublesome features. This example is armed with a 76mm M1A1 gun and fitted with HVSS suspension. *(TM)*

shipbuilding programmes to a massive expansion of the aviation industry while simultaneously, although largely unknown at the time, the Manhattan Project was expanding to develop and produce the Atomic Bomb. The Manhattan Project alone absorbed the equivalent of, collectively, the national wealth of many other nations.

Yet even within such an industrial framework the M4 production programme was still outstanding. Between 1942 and 1945, US industry managed to churn out 33,671 of the M4 armed with the 75mm gun, 18,705 of which were delivered on Lend-Lease terms. It has to be emphasised that those totals relate just to the 75mm gun M4s. To this total must be added a further 10,883 M4s armed with 76mm guns and a further 4,680 mounting 105mm howitzers. The latter two armament categories form the main content of this account yet they cannot be described alone; they have to be seen as part of the overall Medium Tank M4 story.

There are added complications that influence this account. Despite the huge numbers of M4s manufactured and shipped to the various combat theatres there were rarely enough to meet the demands of the front line units at any one time. This led to all manner of field improvisations and expedients such as the removal of equipment from one damaged or unserviceable model of M4 to replace that of an entirely different model. In addition, turrets could be readily swapped between models. Some of these improvisations formed the basis for official programmes conducted in local field workshops and included the incorporation of improvements and enhancements covering every aspect of tank technology. To these measures could be added the rebuilds of combat-weary M4s carried out at repair bases which often meant that what was once a 75mm gun M4 became a 76mm gun M4 (or vice versa), with all the various detail differences between models that could result. When design improvements were introduced to the suspension and other aspects of the production M4, the model discrimination picture became even more confusing but a general outline will be provided in the appropriate text section, where possible.

Above: Developed at around the same time as the Medium Tank M4 series, the Heavy Tank M6 was armed with a 3-inch (76.2mm) gun. For a short while it was the heaviest and most powerfully armed tank in the world but it proved to be too heavy and complicated (among many other short-comings) and only 40 were built. This is the M6A1 with an all-welded hull. It was to have been the main production model. *(TM)*

Right:
An early production example of a Medium Tank M4A1, clearly showing the rounded contours of the cast hull and turret. The cast hull was unique to this model due to difficulties in manufacturing cast components this large. The M4A1 was the first of the M4 series to enter mass production at the Lima Locomotive Works during February 1942. *(TM)*

Below:
On active duty somewhere in North Africa during a refuelling stop, these British Shermans are probably all M4A2 models, the British Sherman III. *(TM)*

Left:
On public display on Horse Guards Parade in London, this M4A1 (Sherman II), the first to reach Great Britain in November 1942, is fitted with the 75mm M2 gun. The barrel stabilisation system was installed as for the M3 gun, the M2 barrel having to be fitted with a counter-weight. Also still installed are the twin machine guns in the front hull. The latter were soon removed as they were of little combat use. *(TM)*

Below:
A New Zealand Army field workshop somewhere in the Western Desert. Four Shermans are visible. *(TM)*

These mentions of variations are placed here to emphasise that there was no hard and fast standardisation across the M4 range of models. What follows can only be an overall guide. There were numerous deviations from the official standardisation intentions so the M4 became a happy hunting ground for the armoured vehicle historian and enthusiast, especially when the number of variations was greatly enlarged by the various M4 variants that were developed, tested or introduced into combat. These varied from tank destroyers to mine-clearers and from bridgelayers to armoured recovery vehicles. We must therefore concentrate only on the up-gunned M4s and Shermans produced during the World War Two years.

As the production totals mentioned above indicate, the most commonly encountered main armament on the M4 was the 75mm M3 gun, the original design of which could be traced back to the French mle 1897. This proved to be a successful and dependable all-round tank gun with the added advantage that it fired high-explosive shells in addition to the armour-piercing projectiles that were the main anti-armour ammunition. But by 1943 the German Tiger and the Panther were in the offing, both having guns more powerful and with longer combat engagement ranges than the 75mm M3. The Tiger's armament was a potent 8.8cm gun while the Panther had a long-barrelled, high-velocity 7.5cm gun. The immediate US response was to consider a variant of the existing 3in (76.2mm) anti-aircraft gun, the M7, originally intended for the Heavy Tank T1 (later type - classified as the Heavy Tank M6 - only 40 were built). It soon transpired that the M7 gun would be too heavy and bulky for the M4 so a new high-velocity gun was developed, the 76mm Gun M1.

Preparing the M4 tank to carry the new 76mm gun was not a straightforward process for the overall dimensions of the new gun were such that it could not be accommodated inside a standard M4 turret as the recoil length was too great. This difficulty was overcome by adopting and modifying the turret originally intended for the Medium Tank T20 and T23. Both the T20 and T23 series of tanks were ultimately cancelled, although they formed part of the development path to the Heavy Tank M26 Pershing. They were both designed to carry the 76mm M1 gun (among other armament options) from the outset, so the turret transfer saved considerable development time and resources. From February 1944 onwards the new gun and turret were introduced onto the M4 production lines. It should be stressed that the arrival of the new gun did not immediately lead to the withdrawal of the 75mm gun M4s. Many of them remained in front line service until (and after) the end of hostilities in 1945.

As the British Army was unlikely to receive 76mm gun M4s in time for the planned invasion of Europe during 1944 it had to develop and produce its own equivalent. (The first deliveries of 76mm gun M4/Shermans for the British Army were not made until December 1944.) This equivalent was intended to be the Cruiser Tank, Challenger (A30), armed with a tank

version of the towed 17-pounder anti-tank gun having a calibre of 76.2mm. As an insurance against the Challenger proving unsatisfactory, the identical main armament was proposed for the Sherman, then in British service in increasing numbers. It was just as well, for getting the Challenger into production took far more time than had been foreseen, to the extent that it was not yet ready when the invasion took place on 6 June 1944. The safeguard 17-pounder Sherman then came to the rescue and conversions from existing Sherman tanks were ordered from February 1944 onwards. The required quantities for the invasion were not met as supplies of the 17-pounder were delivered at a slower rate than had been planned, partially due to urgent demands for the gun in its towed form. The 17-pounder Sherman, known as the Firefly, therefore had to be issued on a limited basis, usually at the rate of one per tank troop. The British then had their own solution to tackling the German Tiger and Panther tanks.

The other main armament change for the M4 series was to install a 105mm Howitzer M4 in a largely unaltered M4 tank turret. The intention here was to produce a close-support tank that could provide direct and indirect fire support for standard-gunned M4 tanks during operations. More details are provided in the appropriate text section.

Above:
An M4A1, believed to be of the US Marine Corps, named *Shanghai-Lill*, lumbering ashore on North New Guinea during late 1942, complete with an infantry squad hitching a ride. (*TM*)

CHAPTER TWO

M4 PRODUCTION

M4 production totals ran into tens of thousands. Starting from a newly established base during 1942, M4s soon poured from the production lines established at several locations in the USA. Two of the main production centres were built specifically for producing tanks while others were more used to building automobiles or locomotives.

Between the wars the small amount of US armoured combat vehicle production that did take place was largely centred around the Rock Island Arsenal in Illinois. When large-scale tank production plans were prepared during late 1940 it was appreciated that Rock Island would be unable to manufacture the quantities anticipated. Although it had the necessary heavy machinery to manufacture tanks, it was primarily an ordnance arsenal specialising in recoil mechanisms. Artillery, from light to heavy, was soon to be as much in demand as tanks. Some other tank manufacturing resource was needed.

Two measures were introduced. One was the construction of a dedicated tank manufacturing plant on a green field site close to Detroit, Michigan. Known as the Detroit Tank Arsenal, the entire facility was run by one of the concerns involved in the second production measure - the use of US commercial industry. Until 1940 military equipment production was the remit of the US Government and Government-owned facilities. Commercial concerns participated in military-related production only rarely. The unprecedented demands of World War Two soon brought about a rapid change.

At first it seemed that the locomotive and associated industries would be sufficient for the task as they had the heavy engineering know-how and industrial plant required to manufacture tanks. However, once the projected numbers began to soar it became necessary to involve the US automotive industry as well. It responded with a will and directed its considerable knowledge of mass production techniques towards tank production. As this account will reveal, nearly all the major US automotive manufacturers became involved in the Medium Tank M4 series production programme. For instance, the Chrysler Corporation managed and ran the Detroit Tank Arsenal on a day-to-day basis.

Some of this preparation of commercial concerns to become involved in armoured vehicle production had been partially primed by the British. Desperately short of all types of military equipment following the Dunkirk evacuation of June 1940, they repeated the procedures they had adopted during World War One by contracting US firms to at least partially fill some of the gaps in their armoury. As there was no way that American industry could spare any capacity to manufacture British tank designs,

(Cont p23)

Above:
M4s and Fireflys of the Canadian Fifth Armored Division on their Victory Parade at Elde Airport, Holland, on 23 May 1945. For parade effect, M4s with 75mm guns are interspersed with longer-barrelled 17-pounder gun Fireflys. *(TM)*

Left:
An Australian Army M4A3 (Sherman IV) operating in New Guinea during early 1945. This example has appliqué armour on the side and a rack for spare track shoes on the front hull. *(TM)*

Above:
A parade ground example of a British Army Sherman I (M4) with the 75mm gun mounted in the Gun Mount M34 - the later M34A1 had a wider gun shield. *(TM)*

Right:
A Sherman III (M4A2) about to wade ashore onto an Italian beach in late 1943. It is a command vehicle, well laden with combat accessories and with a wading trunk over the hull rear. *(TM)*

Above:
Canadian Army Shermans preparing for combat 'somewhere in Italy'. The vehicles appear to be M4A2s armed with 75mm guns. Note the stacks of ammunition and containers. *(TM)*

Left:
Photographed in Burma, this is believed to be a Sherman III (M4A2) of the 14th Army, complete with items stowed on the hull front. *(TM)*

Right:
At first sight this photograph shows a 75mm Gun M3 in Gun Mount M34A1 being changed for a new item. Closer examination shows it is more probably a salvage team at work as an anti-tank round has punched a hole through the appliqué armour plate on the hull side of the tank. *(TM)*

Above:
The first in a Free French sequence of photographs, this M4A2 has extra appliqué armour over the front of the turret to increase protection. A total of 755 of the M4 series tanks were handed over to the Free French as part of the Lend-Lease Program. *(TM)*

Right:
Reloading a Free French Medium Tank M4A2 with high-explosive ammunition, indicated by the stand-off impact nose fuses. This is a training range photograph for, as a safety measure, a red flag is indicating that the vehicle carries live ammunition. *(TM)*

Above: Preparing for 'The Return', Free French M4A2s undergoing live firing training during May 1944. Note that the third vehicle from the front is equipped with dust shields over the tracks. *(TM)*

they had to accept existing or planned US designs which, relating to our account and in the immediate term, initially meant the Medium Tank M3, soon known to the British Army as the Grant.

Among the concerns involved with the British M3 contract (the cost of which was a major contributing factor to the bankruptcy of the British Treasury soon after) were the Baldwin Locomotive Works, Lima Locomotive Works, Pullman Standard Car Company and the Pressed Steel Car Company. These were joined by copious sub-system manufacturers, typically the Mack Manufacturing Company (transmissions) and the Continental Motors Corporation and Wright Aeronautical Corporation (engines). All these concerns later become involved in the M4 production saga.

By the time the M4 entered series production during late 1941 the above-mentioned industrial concerns had been joined by the Pacific Car and Foundry Company, the Fisher Division of General Motors, the Ford Motors Company, the Montreal Locomotive Works in Canada, and the Federal Welder and Machine Company. By early 1942 there were thus 11 major industrial concerns earmarked or actively involved in the final assembly stage of the Medium Tank M4 programme.

They were joined from July 1942 onwards by a second purpose-built Tank Arsenal at Grand Blanc, Michigan, this time run by the Fisher Body Division of General Motors and so generally known as the Fisher Tank Arsenal. It was designed and equipped from the outset for Medium Tank M4 production. As the construction of the plant had commenced in January 1942, from open site to first production tanks rolling off the lines took just six months.

By the end of 1942 the M4 production situation had stabilised to the point where more realistic requirement forecasts could be made. The requirement for 1943 was set at 24,582, plus a further 23,595 during 1944. Needless to say the military situation at any one stage of the war could affect these requirements. For instance during late 1943 the total requirement for that year was altered down to 21,404. The actual 1943 production total was 21,245. By late 1944 it was possible to issue a requirement for 13,574 M4s in anticipation of production emphasis being switched to the Heavy Tank M26 General Pershing. In January 1945 this was raised slightly to 'about 14,500' M4s in response to the late December 1944 German

Ardennes Offensive which seemed to indicate that the war might continue longer than had been hoped. It was also forecast that 'about 3,500' M4s would be required during 1946. Both requirement totals became hypothetical when the cessation of M4 production was ordered and completed by July 1945.

Actual production totals for all Medium Tank M4 production, by year, were as follows:

1942	1943	1944	1945	Total
8,017	21,245	13,179	6,793	49,234

These totals can be broken down into the main models of M4, starting with the M4 tanks with 75mm guns:

1942	1943	1944	1945	Total
8,017	21,245	3,758	651	33,671

Production of the 76mm gun and 105mm howitzer M4 models, the main subjects of this account, did not commence until 1944. Their totals were as follows:

76mm	1944	1945	Total
	7,135	3,748	10,883

105mm	1944	1945	Total
	2,286	2,394	4,680

To place these totals into perspective, the final total of all types of tank - light, medium and heavy - manufactured in the US between 1940 and 1945 was 88,276. During the same period the Germans manufactured 24,360 armoured vehicles of all kinds. The British manufactured 24,803.

Lend-Lease M4 series totals were also important in the production context as the final Lend-Lease allocation reached 22,098.

The total number of 75mm gun M4 tanks handed out on Lend-Lease was 18,075. The main recipients were Britain and its Commonwealth allies who received 15,256. Canada received just four M4s, the Free French 755, while 2,007 were shipped to the Soviet Union. An undefined number of 'American Republics', probably mainly Brazil, received 53 75mm gun M4s.

Lend-Lease issue of the 76mm gun M4 could not begin until late 1944 and only after the immediate needs of the US armed forces had been met but, even so, the Lend-Lease total reached 3,430. Of these, 1,335 went to the British and 2,095 to the Soviet Union.

Above:
Photographs of Lend-Lease vehicles in service with the Soviet Union forces are rare. This is of an all-steel tracked M4A2 shows several local additions such as fuel drums all over the vehicle and an unditching beam. There is also the usual complement of 'tank descent' infantry. *(TM)*

Far left:
September 1944: British infantry and armour advance through the Appenines, Italy, with a British Army Sherman OP (observation post) or headquarters tank. Note the extra antennae and a telephone cable drum secured to the hull front. *(MP)*

Above:
M4A4s of the Chinese-American 1st Provisional Tank Group operating in Burma during early 1945. *(TM)*

The only Lend-Lease recipient of the 105mm howitzer M4 was the British Army. They received 593, although it is possible that not all of these were delivered as the allotment was made late in the war.

To continue the statistical theme, a summary of the M4 totals (of all types) achieved by individual manufacturers during 1943 and 1944, the 'big production' years, was as follows:

	1943	**1944**
American Locomotive	2,174	-
Baldwin Locomotive	1,100	43
Detroit Tank Arsenal	6,612	5,587
Federal Machine	519	-
Fisher Tank Arsenal	2,240	6,052
Ford Motor	1,176	-
Lima Locomotive	835	-
Pressed Steel Car	3,000	2,171
Pacific Car & Foundry	660	-
Pullman Standard	3,003	-

(Note: There is a discrepancy between official [on the previous page] and manufacturer's production figures. This is thought to have occurred when chassis were diverted for experimental use.)

Unfortunately, accurate corresponding figures for 1942 and 1945 have not been found. It will be noted that by the end of 1944 the main M4 production centres had moved away from the industrial concerns and were increasingly concentrated on the two Tank Arsenals. The industrial concerns were then free to concentrate on manufacturing other types of armoured vehicle, including the Heavy Tank M26 that was intended to gradually remove the main armoured combat burden from the M4.

While using the production figures outlined above it would be apposite to compare them with those for that other war-winner, the Soviet T-34 series. Unfortunately, precise totals do not seem to be available and the following are only educated estimates:

1940	**1941**	**1942**	**1943**	**1944**	**1945**
115	2,810	ca5,000	ca10,000	11,758	ca10,000
				Total:	40,000

To round off this number-crunching, by 1952 US Government accountants had taken the considerable trouble to work out the unit cost of each type of US armoured vehicle manufactured between 1942 and 1945. Although indicative rather than accurate, the results for just the Medium Tank M4A3 were as follows:

Medium Tank M4A3, 75mm gun	US$47,339
Medium Tank M4A3, 76mm gun	US$55,145
Medium Tank M4A3, 105mm howitzer	US$52,929

Above:
Inside a REME depot workshop at Aldershot during September 1944. The example in the foreground appears to be a Sherman III (M4A2). The third Sherman from the right is equipped with a mine flail. *(TM)*

Left:
A typical REME depot workshop task: lifting the turret from a Sherman I (M4). The drive sprockets at the front have both been removed. Note also the spare track shoe rack on the hull front . *(TM)*

CHAPTER THREE

M4 BASE MODELS

The rapid expansion in production demands for the Medium Tank M4 meant that throughout its life engines other than the intended air-cooled radials had to be utilised. This led to five base models of the Medium Tank M4, with four possible engine installations.

When the US medium tank programme began in earnest during the late 1930s it was decided that for reasons of economy and to save development time the main power plant would be an aircraft air-cooled radial engine. In addition, using such an engine would result in the provision of the necessary horsepower from a relatively compact and reliable unit. However, there was a snag, one that had already been highlighted during the hurried mass production of the Medium Tank M3. As M3 production began the US aviation industry was faced with an identical capacity expansion challenge. Aircraft engines soon became a production bottleneck as they were urgently needed for aircraft, especially training aircraft.

Alternative engine installations therefore had to be devised and utilised, as had already happened with the Medium Tank M3. Thus at one time the M4 series could have been fielded with five different engine installations, although this number was soon reduced to four in an attempt to somehow rationalise matters. Only one of these four engines, the Ford GAA, was developed specifically for tanks.

Another major difference between M4 models was the nature of the armour employed and its arrangement. At one early production stage the supply of suitable armour threatened to be another major bottleneck. As with many other military commodities, the demands that were about to be imposed on US industry to supply the main types (face-hardened, cast and rolled homogeneous) had not been foreseen. Only strenuous efforts by the US steel industry, soon working flat out to meet the many demands on its products, managed to overcome that supply problem. Among the many urgent measures taken was the establishment of the Gary Armor Plate Plant in Illinois, itself a major undertaking equivalent to that for the two Tank Arsenals. By the time the M4 was in production the use of face-hardened armour plate was no longer favoured as welding and casting became the preferred production techniques, with casting offering the added protection factor of smooth-contoured surfaces and edges to deflect incoming shot. Of the two, body castings were preferred to welded equivalents although a simplified welded body had to be used for all but one model to ensure production could be carried out in the required numbers. Relatively few US foundries could cast the large hull and turret assemblies to the

Above:
Inter-allied co-operation with a US Army Ward La France Series 2 Heavy Wrecker assisting in a Wright Continental R-975 C1 engine change on a British Army Sherman I (M4). Compared with many other contemporary tank designs, this was a comparatively rapid and easy procedure on a Sherman. *(TM)*

Left:
US Army fitters carrying out final checks on the waterproofing of a line of what are almost certainly M4A3 models during early June 1944. Note the extra armour added to the turret fronts. *(TM)*

Right:
The Chrysler A-57 Multibank petrol engine as fitted to the Medium Tank M4A4 (Sherman V). This is the left-hand side of the engine with the drive shaft pointing towards the transmission unit mounted in the nose of the tank. *(TM)*

Right:
The five banks of the Chrysler A-57 Multibank engine, making 30 cylinders in all, can be appreciated from this frontal view. Each bank was formed from a Chrysler automobile engine. *(TM)*

Left:
The power plant for the Medium Tank M4A2 was formed by combining two General Motors 6-71 diesel engines, normally used to power trucks, in a side-by-side configuration so that they had a common drive shaft. *(TM)*

Left:
This front-on view of the power plant for the Medium Tank M4A2 demonstrates how each of the two six-cylinder General Motors 6-71 diesel engine outputs were arranged to power a common drive shaft. *(TM)*

Right:
The Wright Continental R-975 C1 air-cooled radial engine was originally designed as an aircraft engine but was easily modified for powering tanks. It was used to power the Medium Tank M4 and M4A1 but shortages of this engine meant that other engines also had to be employed.

extent necessary so just the one M4 model, the M4A1, had the cast hull.

As mentioned elsewhere, in practice there was no clear differentiation between all the models mentioned below as there were many other factors involved, apart from engines and armour. One was the suspension that changed from vertical volute suspension system (VVSS) to horizontal volute (HVSS) on several models during 1944, as will be explained. One further complication was that some models could be encountered with two or three possible armament calibre installations, the M4A3, for instance, being manufactured with three main armament options.

Then came the constant list of modifications that were gradually introduced across the board throughout the M4 production period, ranging from rewiring to more significant measures, one example being the provision of a loader's hatch for the turret. At one stage over 3,000 engineering changes were introduced to the production lines during a single month, either to simplify or speed manufacture or as the result of combat experience.

To add to all these variations there were the rebuild or rework programmes. In the main these involved a drastic refurbishment of tanks employed for training in the USA. When trained troops were sent overseas they were usually issued with what was already in theatre or newly delivered vehicles. Gradually a stock of well-used ex-training vehicles grew to the point where they could swell the numbers being sent overseas. This could follow an extensive programme of stripping down, cleaning and modifying up to the latest production standards, and providing engines with a thorough overhaul or even replacement. Extra armour was usually added as well. In addition to these measures, batches of battle-weary tanks were returned to the US for the same treatment, or underwent similar update and repair programmes at US Army tank depots in Europe and elsewhere.

One result of all this rebuild/rework activity was that all manner of updating measures were

incorporated to the extent that overlaps between base models became blurred. What follows is therefore a brief summary of the base models, including the British designations where appropriate.

Medium Tank M4

This was intended to be the base model for the Medium Tank M4 series. It was powered, as planned, by a de-rated nine-cylinder Wright Continental R-975 C1, air-cooled, radial petrol engine delivering a nominal 400hp at 2,400rpm (these engines were also manufactured by Continental Motors). The hull was welded although the nose was originally constructed from three sections bolted together. Late production examples had a one-piece cast nose or a combined cast/rolled hull front, both measures adding extra protection. Another major change was that the original M34 gun mount with a narrow mantlet was changed to the M34A1 mount with a wider mantlet to give better ptrotection.

The base M4 was manufactured by Pressed Steel (the first of this model entered production in July 1942), Baldwin Locomotive Works, American Locomotive, Pullman Standard Car, and the Detroit Tank Arsenal.

The base M4 was one of the models to be armed with the 105mm Howitzer M4. About 800 were manufactured at the Detroit Tank Arsenal with the vertical volute suspension, followed by a further 841 examples with the horizontal volute suspension system. In both cases the vehicles were known as the M4 (105mm).

For the British, the base M4 was the Sherman I. The Sherman Hybrid I was a late production model built at the Detroit Tank Arsenal with the combination cast/rolled hull front. In British service the M4 (105mm) became the Sherman IB while late examples of the M4 (105mm) Howitzer model with HVSS became the Sherman IBY.

Above:
The Medium Tank M4A1 is the easiest of the M4 series to identify as it was the only model with a cast hull, complete with rounded contours. This example, serving in the Pacific Theatre during late 1944, has track grousers added to improve traction over soft terrain. *(TM)*

Right:
A Wright Continental R-975 C1 engine change for a Sherman II (M4A1) being carried out by REME fitters using a Heavy Wrecker M1A1, a model built by two US companies, Ward La France and Kenworth. *(TM)*

Above:
The long barrel and rounded hull contours mark this example as a Medium Tank M4A1 (76mm). The revised pattern of turret was taken from the Medium Tank T20/T23 series that otherwise had little impact on US tank development. *(TM)*

Right:
Combat in the Bocage country of Normandy after June 1944 introduced a new combat accessory, the Culin Hedgerow Device hedge cutter. The example on the front of this M4A1 (76mm) was known as Rhinoceros. Note also the layer of sandbags on the hull front, intended to provide extra protection against German 8.8cm *Raketenpanzerbüchse* shaped-charge warhead rockets. *(TM)*

Above:
This overhead view of a Medium Tank M4A1 (76mm) clearly shows the revised turret outline made necessary by the bulk and recoil length of the 76mm Gun M1. *(TM)*

Left:
Although it was intended that the M4 76mm gun series would replace all earlier 75mm gun models the change-over was far from complete when the war ended ,as this late-war period M4A1 shows. *(TM)*

Above:
A late production (but still with VVSS) and factory fresh Medium Tank M4A3, powered by a Ford V-8 GAA petrol engine. Note the one-piece cast nose and revised front hull angle. This vehicle is complete with dust shields over the tracks and has shipping covers in place. *(TM)*

Medium Tank M4A1

The M4A1 was essentially the same as the base M4 but with a cast hull, as originally intended. Despite the A1 suffix denoting otherwise the M4A1 was actually the first of the M4 series to enter production at the Lima Locomotive Works in February 1942, Lima being responsible for the first production pilot model derived from the Medium Tank T6 prototype. The first examples were delivered armed with the 75mm Gun M2 that had a barrel shorter than the intended 75mm Gun M3, and had spoked road wheels, both the gun and wheels being carried over from the Medium Tank M3 production lines. Lima was joined in production by Pressed Steel (from March 1942) and the Pacific Car and Foundry Company (from June 1942).

The M4A1 was one of the models to be armed with the 76mm Gun M1. A total of 3,396 were manufactured by Pressed Steel and known as the M4A1 (76mm).

In British Army service the M4A1 became the Sherman II. Examples of the M4A1 (76mm) became the Sherman IIA, while the Sherman IIC (Firefly) was a M4A1 rearmed in the UK with a 17-pounder gun.

The M4A1 was also produced in Canada as the Cruiser Tank, Grizzly. A small number, with some minor equipment changes to suit Canadian armed forces needs, were manufactured by the Montreal Locomotive Works between September and December 1943. The production line then switched to producing the Sexton self-propelled gun carriage armed with a 25-pounder field gun. By late 1943 this was considered a better use for the Montreal Locomotive Works facilities rather than merely supplementing the massive US Medium Tank M4 programme, by then producing all the M4s likely to be needed.

At one stage the US Ordnance Department modified a M4A1 to accept a Ford GAA engine

Above: This overhead photograph shows an early production Medium Tank M4A2 with tools and other equipment stowage displayed in military manual fashion. *(TM)*

as a feasibility study. The combination, the M4E7, was not proceeded with.

Medium Tank M4A2

The M4A2 followed much the same welded hull lines as the base M4 and was the second M4 model to enter production. It was the first M4 model to enter production at the newly constructed Fisher Tank Arsenal in April 1942. Also involved in M4A2 production were Pullman Standard Car, American Locomotive Works, Baldwin Locomotive Works (although it manufactured only 12), and the Federal Welder and Machine Company.

The main significance of the M4A2 was that it employed an engine pack formed by connecting two six-cylinder, General Motors 6-71 6046 liquid-cooled in-line diesel engines in a side-by-side configuration to a common prop shaft. Delivering, in combination, a nominal 410hp at 2,100rpm, these engines were originally developed for trucks but proved highly successful and reliable in the M4A2, even if drivers often found it difficult to keep the two engines synchronised at times.

Almost the entire M4A2 production output, other than a batch for the US Marine Corps, was passed over to the Lend-Lease programme, the main recipients being the UK and the Soviet Union. The Soviets preferred this model to other M4 models as the diesel engines simplified their front line fuel supply logistics; they knew it as the M4A2 ШЕРМАН (Sherman).

The M4A2 was one of the models that could be armed with the 76mm Gun M1. In total 3,396 were manufactured by the Fisher Tank Arsenal and Pressed Steel. This model remained in production by Pressed Steel until June 1945, known as the M4A1 (76mm).

For the British Army the M4A2 became one of its mainstays. It was known as the Sherman III, examples with the 76mm gun

COIFFEUR

being the Sherman IIIAY. The latter were delivered only from late 1944 onward.

Medium Tank M4A3

Apart from the welded hull, the main feature of the M4A3 was the engine, the only one in the M4 power plant series designed specifically to power tanks. It was the Ford V-8 GAA, liquid-cooled petrol engine delivering a nominal 500hp at 2,600rpm. This engine proved to be so reliable and trouble free that the M4A3 rapidly became the preferred US Army model, none being passed to the Lend-Lease programme until late 1944 and only after the requirements of the US armed forces had been met.

As well as supplying the engine, Ford Motors manufactured 1,690 M4A3s between June 1942 and September 1943. During September 1943 M4A3 production was relocated to the Fisher Tank Arsenal and continued there until March 1945.

The Medium Tank M4A3 underwent the most production, modification and armament changes of any of the M4 series. The most significant were the introduction of the horizontal volute suspension system (HVSS) and changes to the main armament as both the 76mm Gun M1 and 105mm Howitzer M4 were involved with this model. These armament changes were denoted by M4A3 (76mm) and M4A3 (105mm). There were also the M4A3 (76mm) HVSS and M4A3 (105mm) HVSS.

There was also an M4A2E2, categorised as an Assault Tank. This was a rushed measure devised from early 1944 onwards when it was realised that the Heavy Tank M26 would not be ready for service until early 1945. As a heavily armoured tank was required to provide close support for infantry operations it was decided to add extra hull and turret armour to the M4A3. Hull armour was 3.94in (100mm) thick while the turret was an entirely new box-contoured component with frontal armour 5.91in (150mm) thick. As these measures increased the vehicle weight considerably (to 40.69 tons [41,342.73kg]), wide all-steel tracks with grousers as a permanent fixture were employed. The one anomaly was that the 75mm Gun M3 was

Above:
A standard Medium Tank M4A2 complete with dust shields over the tracks, appliqué armour over the vulnerable ammunition stowage points on the hull sides and the turret top-mounted Browning 0.50-calibre M2 air-defence machine gun. *(TM)*

Far left:
A Medium Tank M4A2 working through the approaches to Brest, France, in late July 1944. As it had had to travel its way through the dense Bocage country of Normandy the Culin Prong is still mounted on the front hull. *(TM)*

Above:
Top view of an M4A3E2, an up-armoured variant of the M4A3 intended to act as a heavy 'assault tank'. Note the extra layer of armour added around the turret sides and the stowage of the 0.50-calibre M2 heavy machine gun (with barrel removed) at the rear of the turret. *(TM)*

Right
Rear view of a Medium Tank M4A3 (76mm) complete with dust shields and HVSS but without most of the normal tools and other stowed equipment. *(TM)*

retained. Once in service in Europe some of these tanks, widely known as the Jumbo due to their reduced speed and manoeuvrability, were rearmed with 76mm guns and turrets taken from damaged M4 (76mm) tanks. The Fisher Tank Arsenal produced 254 new M4A3E8s during May and June 1944. As soon as they were completed they were rushed to Europe, the only theatre in which they served, but only as 'limited standard' tanks. They were withdrawn as soon as the Heavy Tank M26 became available.

Although they did not receive any examples until late 1944, the British Army designated the M4A3 as the Sherman IV. The M4A3 (76mm) became the Sherman IVA and the M4A3 (105mm) became the Sherman IVB. The Sherman IVC (Firefly) was a M4A3 rearmed in the UK with a 17-pounder gun.

Medium Tank M4A4

The main feature of the M4A4 was the engine installation, another expedient to overcome the radial engine shortage and one carried over from the Medium Tank M3 series. It was the Chrysler A-57 Multibank petrol engine, a remarkable improvisation produced by combining five six-cylinder Chrysler production automobile petrol engines, forming a total of 30 cylinders, located around a central drive shaft to deliver a nominal 370hp at 2,400rpm. The result was successful but required a larger engine bay than the other M4 power units. The M4A4 hull was therefore lengthened to provide the extra space and the suspension units were relocated accordingly. This length increase also resulted in a longer track. The M4A4 was manufactured only at the Detroit Tank Arsenal from July 1942 onwards, the final total in September 1943 being 7,499. The M4A4 was the first of the M4 series to be phased out of production and retained the 75mm Gun M3 throughout its production life. It also retained the welded hull of the standard M4.

To the British, the M4A4 was the Sherman V. This model was the main one used for the Firefly conversion involving the installation of a 17-pounder gun, the result becoming the Sherman VC (Firefly).

Above:
The same M4A3 (76mm) vehicle as shown (far left below). Note the long 76mm gun barrel overhang, HVSS, and the revised turret outline needed to accommodate the longer and heavier gun. *(TM)*

Above:
The pilot vehicle for the Medium Tank M4A4 (Sherman V) series, powered by the Chrysler Multibank engine. The 75mm gun is in the narrow Gun Mount M34 and the tank is fitted with the three-piece nose. *(TM)*

Right:
Although it was visually not much different from other all-welded models in the Medium Tank M4 series, the M4A4 (Sherman V) was actually longer than the other models to accommodate the bulk of the Multibank engine. *(TM)*

Above:
The Cruiser Tank, Grizzly I, manufactured by the Montreal Locomotive Works, was the Canadian version of the M4A1 and differed only in detail from the US original. A total of 188 were built. *(TM)*

Left:
An M4A1 passes through an English town during pre-D-Day exercises. *(TM)*

Above:
The M4E5 used as the trials vehicle for the M4 (105mm) howitzer series, early 1943. *(TM)*

Right:
A column of M4A3 (105mm) fire support tanks of the US 3rd Army moving up to positions near Bitburg, February 1945. *(TM)*

Far right above:
Pilot model of the M4 (105mm), M4E5, ready for trials at Aberdeen Proving Ground, early 1943. *(TM)*

Far right below:
A late production M4 (105mm) fire support tank complete with HVSS, sand shields and barrel clamp. This model was known to the British Army as the Sherman IB. *(TM)*

M4
11435
ARMORED BOARD
TEST OPERATION

T-270838

Right:
Demonstrating the US Army's disregard for the Nazi national symbol, this Medium Tank M4 series 75mm gun vehicle is difficult to identify further as the entire front hull is hidden under a layer of cement. *(TM)*

During early 1943 two examples of the M4A4 were used as pilot models for the M4 (105mm), redesignated as the M4A4E1. In the event no further 105mm Howitzer versions of the M4A4 were produced.

Medium Tank M4A5

The M4A5 designation was a paper designation only as it was applied by the US Ordnance Department to the Canadian Cruiser Tank, Ram. Although it was a Canadian development, the Ram was based around Medium Tank M3 components such as the suspension, lower hull and drive train. It is believed that the M4A5 designation was issued when the Ram was supplied under the Lend-Lease programme after Ram production had commenced at the Montreal Locomotive Works, a subsidiary of the American Locomotive Works and therefore originally intended to be one of the concerns selected by the US for M4A1 tank production. The 1,899 Rams manufactured were largely funded by the Lend-Lease programme. One Ram was procured by the US Government for trials and comparison purposes.

Medium Tank M4A6

The M4A6 was the shortest-lived of the main M4 models as it was powered by yet another engine expedient, this time the nine-cylinder Caterpillar RD-1820 air-cooled radial diesel engine, also known as the Ordnance Engine. This engine developed a nominal 450hp at 3,000rpm and was housed in the same lengthened welded hull as the M4A4, the interim Medium Tank M4E1 being used to test such an arrangement. Unfortunately the Ordnance Engine was only fitted in the M4A6, the sole US tank-type to be so powered. Production at the Detroit Tank Arsenal followed on from that of the M4A4 in October 1943. M4A6 production was cancelled soon after (in February 1944), by which time only 75 had been completed, the main reason being that the introduction of a diesel radial to the M4 family was one engine option too many.

As far as is known no M4A6 tanks left the US. To prepare for any that might make the journey, the British Army issued the designation of Sherman VII. It was never applied.

CHAPTER FOUR

DESCRIPTION

The Medium Tank M4 series was built in vast numbers and with many detail differences between the various models. These variations were compounded by the addition of battlefield modifications, cannibalisations, refits and refurbishments of varying levels.

When considering US tank designs it has to be stated that the accent was placed on mobility and firepower, in that order. The US Army's tank designers at Rock Island Arsenal placed the third part of armoured combat vehicle design, protection, as the lowest of its priorities but did not totally ignore it. It simply considered protection as less important than firepower and mobility. These were design factors that often translated in combat to mean that the M4 all too frequently suffered badly from incoming fire. What had been good enough protection in 1941 was no longer viable by 1944. Importance was also placed on a further priority, namely mass production.

This distribution of design priorities meant that the lack of proper armoured protection all too often placed the Medium Tank M4 series at a disadvantage when attempting to engage the better-armed and protected German tanks, notably the Panther and Tiger and their adjuncts, the tank-killing Jagdpanzer. Where they displayed an advantage was in having a good level of speed and manoeuvrability, and in being available in large numbers, so large that at times they were often able to overwhelm their adversaries by sheer weight of numbers. The M4 series also had the advantage that, in armoured warfare terms, if one was lost it could be hard on the crew concerned but another M4 could soon take its place. If a German tank was lost it usually proved difficult to replace, especially during 1944 and 1945.

Details

All US models of the Medium Tank M4 had a crew of five: commander, driver, co-driver, gunner and loader. The driver was seated in the hull left front (looking forward) with the transmission housing, connected via a prop shaft to the engine at the rear, to his right. The driver used two levers for steering, the gearbox providing five forward gears and one reverse.

Seated at the hull right front was the co-driver who controlled the hull machine gun. On early models the driver and co-driver were both provided with horizontal vision slits set into the hull. These proved

Above:
A late production example of a Medium Tank M4 operating in the fire support role for the Canadian Army in Italy. *(TM)*

Left:
A well-camouflaged Canadian Army Firefly (Sherman IIC) negotiating a Bailey Bridge erected across the River Santerno, Italy, during 1945. Note the wide grousers (Spuds) added to the tracks. *(TM)*

Right:
Driver controls on an M4/M4A1: steering and braking control was by levers. The driver was provided with an accelerator pedal and also a clutch pedal. There was also a hand throttle (upper right) for cold starting. On the left is the instrument panel and gyro compass. The driving controls were duplicated for the co-driver/hull gunner.

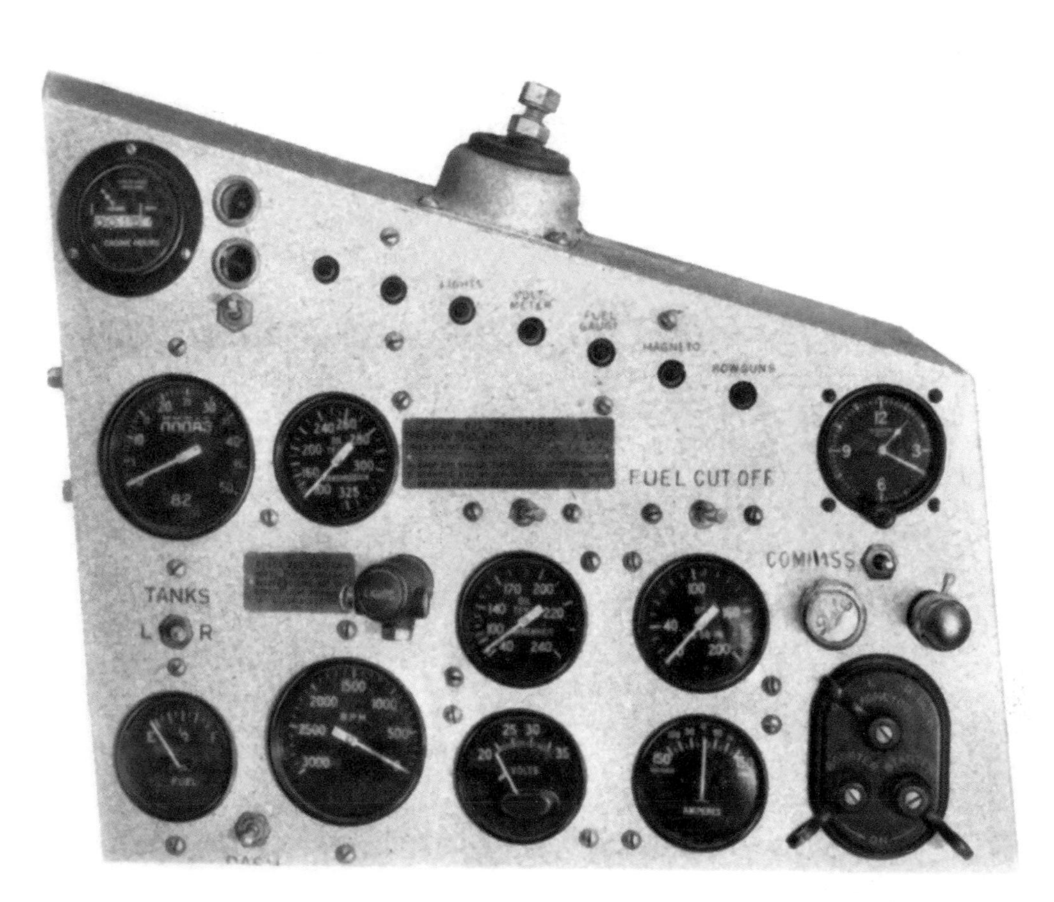

Left:
Close up of the driver's control panel on early models of M4/M4A1.

Left:
The improved driver's control panel as fitted to the Medium Tank M4A1 and later models.

Above:
The rounded contours denote that this example is a Medium Tank M4A1 but the long gun barrel denotes that it is an M4A1 (76mm), known in the British Army as the Sherman IIA. *(TM)*

Right:
The turret for the M4A1 (76mm), taken from the abortive Medium Tank T20/T23, was fitted to all 76mm gunned M4s. *(TM)*

Above:
This late production M4A1 (76mm) shows the one-piece cast nose, the barrel clamp folded down between the drivers' positions, also turret roof detail. *(TM)*

vunerable to enemy fire so they were plugged and replaced by traversing periscopes set in the hull roof. Both the driver and co-driver had their own access hatch.

In the turret, the commander was at the right rear, provided with his own revolving two-piece access hatch. On early models the commander's main vision device was a single traversing periscope in the roof hatch but by the time the M4 (76mm) and M4 (105mm) tanks were in production this arrangement had been replaced by a new design of cupola with six episcopes providing good all-round vision.

The gunner was positioned in front of the commander, to the right of the main armament, which, in the case of the 76mm gun and 105mm howitzer, took up much of the internal space within the turret. The loader was located to the left of the gun, with a traversing periscope for outside vision. On late production models the loader was provided with an enlarged access hatch in the turret roof.

M4 (76mm) tanks could have armoured hulls that were either cast M4A1 (76mm) or welded M4 (76mm) or M4A3 (76mm). M4 (105mm) tanks used only welded hulls as they were based on M4 and M4A3 tanks.

The armoured hull was originally 2.0in (51mm) thick over the hull front, although this was later increased to 2.50in (63.5mm). Thanks to the angled front hull plate this presented a depth of about 3.94in (100mm) to horizontal incoming fire. Late production M4s had the front hull plate re-angled to a steeper angle of 47°. This not only simplified production but provided a little more internal space and slightly improved the protection factor. It also meant that the driver's and co-driver's access hatches could be enlarged. Side armour was a modest 1.50in (38mm) which many tank personnel considered inadequate. As a result many M4s were provided with extra sheets of appliqué armour welded to the hull sides over the area close to the ammunition stowage racks. On some models more armour was welded to the

Above:
A Sherman IB, the British Army designation for the Medium Tank M4 (105mm). *(TM)*

turret front. This appliqué armour was usually added at field depot level. Once in the field all manner of extra protective measures were applied, especially after the German Panzerfaust and Raketenpanzerbüchse portable infantry anti-tank weapons, with their shaped charge warheads, began to become more prevalent from 1943 onwards. It was soon discovered that the further away from the target armour the shaped charges could be detonated after impact, the less the chance that the shaped charge high-temperature jet could penetrate the hull or turret. This led to emergency field measures such as stacking sandbags or logs of timber all around the hull and carrying lengths of spare tracks slung or welded along the sides. Spare wheels were often carried on the hull front. These field improvisations also gave an extra measure of protection against conventional anti-armour projectiles. They also gave an odd and scruffy appearance to many M4s but such measures were deemed necessary, even if only for morale reasons.

Another field measure introduced to reduce the effects of land mines was to place a layer of sandbags along the floor, both under the driver and co-driver and the main combat area. The floor armour was only 1in (25.4mm) thick at the front, reducing to .5in (12.7mm) under the engine compartment. A floor hatch was provided behind the co-driver as a means of escape, when needed. It could also be used

by infantry to pass messages into the tank while under fire as there was no other communication interface between the infantry and tank crew, other than hammering on a closed hatch.

The engine compartment was at the rear with hinged and louvered top covers allowing good access for maintenance or engine replacement. They also provided good ventilation for engine cooling in the case of the radial engine M4 and M4A1 models (all other M4 series engines were liquid cooled, apart from the short-lived Ordnance Engine diesel radial which was discontinued after it had been installed on only 75 M4A6 tanks). The Ford GAA engine used in the US Army's preferred M4A3 model was a V-8 liquid-cooled petrol (gasoline) engine. This engine was positioned on four brackets in the engine compartment, each bracket having a rubber pad to reduce vibrations.The two front brackets were secured to the engine compartment bulkhead between the engine compartment and main combat compartment, the rear two brackets being located on the engine compartment floor. Fuel was carried in two main tanks, one each side of the engine compartment. A two-section drive shaft connected the engine to the transmission in the hull front.

The tracks were driven by drive sprockets at the front of the vehicle, two each side. Adjusting the idler wheel at the rear of the

Top:
The gearbox mounted to the transmission unit in the cast nose of the M4/M4A1.

Above:
The right rear of the M4/M4A1 series gearbox/ transmission unit - note that the final drive sprocket has been fitted.

Right:
A column of M4A3 (76mm) tanks of a US Army 'All Negro' battalion operating in Germany, March 1945. *(TM)*

A9

Above:
This illustration of an HVSS (Horizontal Volute Suspension System) unit clearly shows the horizontal volute (coiled flat spring). *(TM)*

Right:
VVSS (Vertical Volute Suspension System) as fitted to the M4 series. *(TM)*

Above: This example of a late production Medium Tank M4A3 has the track fitted with the 37in (940mm) wide track grousers (Spuds) for crossing soft terrain. In battlefield conditions not all of the grousers would be fitted (see page 64). *(TM)*

hull altered the track tension. Even as late as 1945 the two original types of track could be encountered, both being interchangeable. The first pattern, carried over from the Medium Tank M3, incorporated rubber blocks but to preserve rubber stocks all-steel tracks were introduced. There were 79 track shoes each side, 83 on the longer-hulled M4A4 and M4A6. It was possible to fit grousers to the outer ends of the track to provide more grip and so assist mobility under difficult terrain conditions. When not installed, these grousers were carried in a special compartment in the hull rear. Track life of over 2,485.6 miles (4,000km) was not uncommon, far more than for many types of tank track.

From late 1943 onwards two new track types started to appear, the T80 (all steel) and the T66 (rubber chevron pattern). These were wider than before (actual width 23.01in [584mm] compared to the earlier 16.5in [419mm]) to provide better mobility over soft terrain plus a smoother ride for the crew. The T80 and T66 tracks could only be used in conjunction with a new type of suspension, the Horizontal Volute Suspension System (HVSS).

Until then there had been three Vertical Volute Suspension System units (VVSS, carried over from Medium Tank M3 production) located each side of the hull. With the VVSS the shocks and variations imparted by ground conditions were largely absorbed by two vertical volute (spiral) springs in each unit which also carried two rubber-tyred road wheels. Two patterns of VVSS unit could be encountered, the earliest have the steel track return roller on the top of the unit. The second carried the return roller on a trailing arm. Early road wheels had spokes so to simplify manufacture they were soon replaced by solid steel wheels with hard rubber rims.

Right:
Fitting an early production M4A1 tank with extended end connectors, known to the British as Spuds, on the outer side of each steel track plate. *(TM)*

Below left:
Close-up detail of how the extended end connectors were held in position. *(TM)*

Below right:
Two individual track extended end connectors (Spuds). *(TM)*

On the HVSS the volute springs were horizontal and each unit had four road wheels (two sets of two). The T66 and T80 track had the track guide located centrally - the earlier narrow tracks had guides on each end of the shoe. With HVSS the track return rollers were secured to the hull sides. The number of HVSS units remained at three each side.

The HVSS had the same advantage as the VVSS in that should a unit become damaged it was only necessary to unbolt and change just each individual unit. It was also possible to change individual road wheels on the HVSS unit without having to remove the entire assembly. Both types of suspension unit had the extra advantage that they occupied no space inside the hull and were generally easier and cheaper to manufacture than other suspension systems, such as torsion bars.

Sand or dust shields over the tops of the tracks were available but were not always installed.

Back inside the hull the radio carried in US tanks was the SCR-508 for communications with other tanks within the unit. Some tanks were completed as command vehicles, the only changes from the norm being the provision of a second radio set, the SCR-506, and the necessary aerial for communications with higher command levels. If all else failed, twenty-four M238 signal flags were provided. Every member of the crew had an intercom plug-in point.

Other combat accessories provided included two fixed fire extinguishers in the engine compartment plus another two, smaller, hand extinguishers in the combat compartment. Also fitted in the tank was a Decontamination Apparatus M2 in case of a chemical agent attack.

Despite all the combat equipment carried inside any M4 series tank there still remained ample internal space for the crew's personal kit and belongings, something gratefully accepted by the many Allied tank crewmen who operated the M4.

Armament details are provided in the following section.

Above:
On late production M4 series vehicles with HVSS the track was of the T66 or T88 type with a central track guide arrangement. *(TM)*

Above left:
Detail close-up of the late production outer final drive sprocket showing how worn or damaged sprockets could be readily changed by the simple use of spanners. *(TM)*

Right:
A Firefly IIC fitted with 37in (940mm) track grousers (Spuds) fitted to each alternate track shoe. The location is the attack across the River Reno at San Alberto, Italy, by the Eighth Army in 1945. *(TM)*

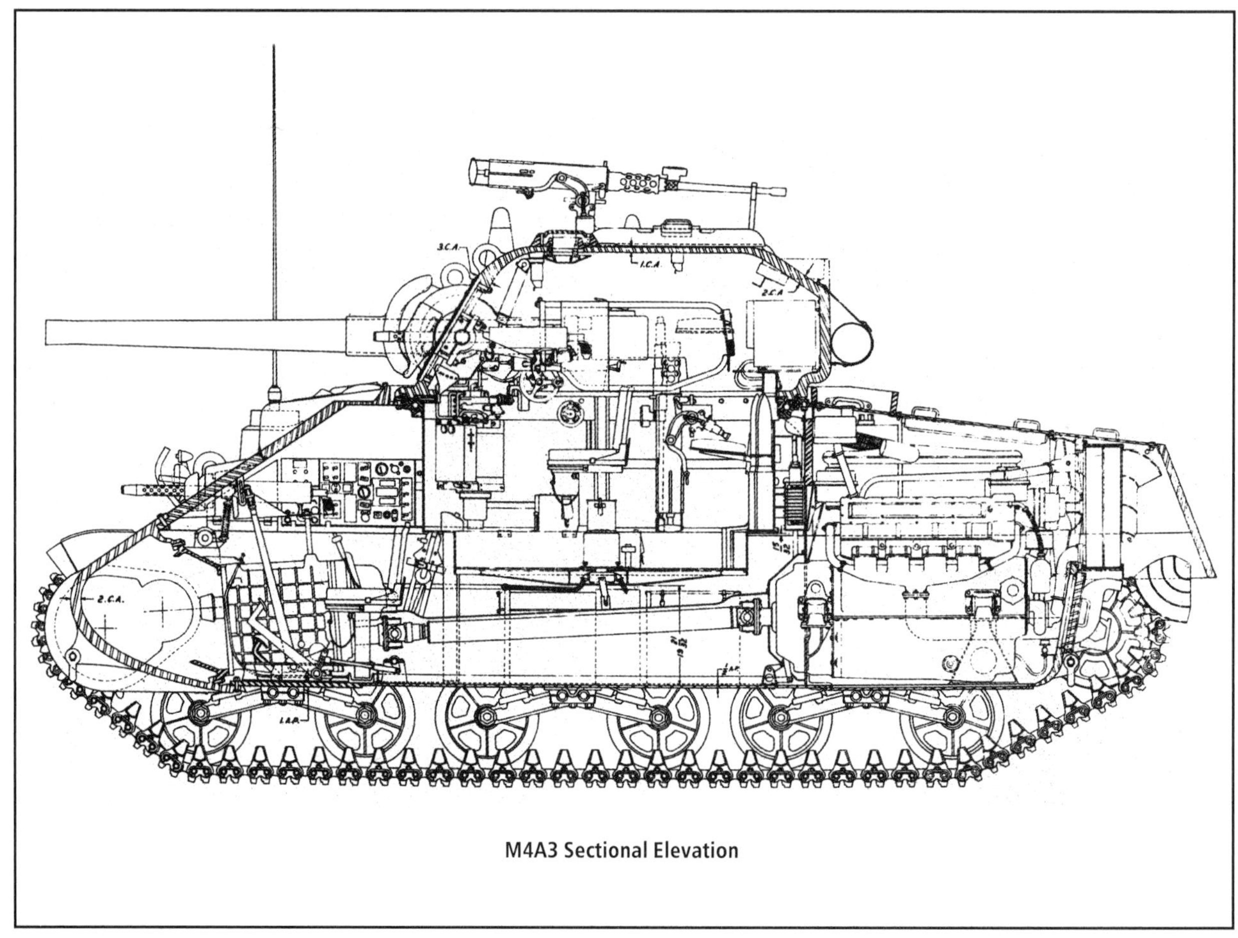

M4A3 Sectional Elevation

Specifications - M4A3 (76mm)

Crew:	Five (commander, driver, co-driver, gunner, loader)
Weight in combat:	ca 31.78 tons (32,284.27kg)
Length, gun forward:	24.26ft (7.39mtr)
Width:	8.79ft (2.68m)
Height, overall:	9.74ft (2.97m)
Track:	6.89ft (2.1m)
Track width:	16.15in or 23.01in (410mm or 584mm)
Ground clearance:	17.14in (435mm)
Max speed, road:	24mph (38.6km/h)
Fuel capacity:	174.9gal (795ltr)
Range, road:	ca 84.5 miles (136km)
Fording:	3ft (914mm)
Vertical obstacle:	2ft (610mm)
Trench crossing width:	7.48ft (2.28m)
Engine:	18,000cc V-8 Ford GAA-III liquid-cooled petrol engine developing 500hp at 2,600rpm
Transmission:	synchromesh, with five forward and one reverse gears
Steering:	Cletrac
Suspension:	VVSS or HVSS
Tyres:	VVSS, 20 x 9
Electrical system:	24V
Armour, front:	2in (50.8mm)
side and rear:	1.5in (38mm)
floor:	max, 1in (25.4mm)
Armament:	1 x 76mm M1 series gun 2 x 0.30in M1919A4 machine guns 1 x 12.7mm M2 HB machine gun 1 x 2in smoke mortar

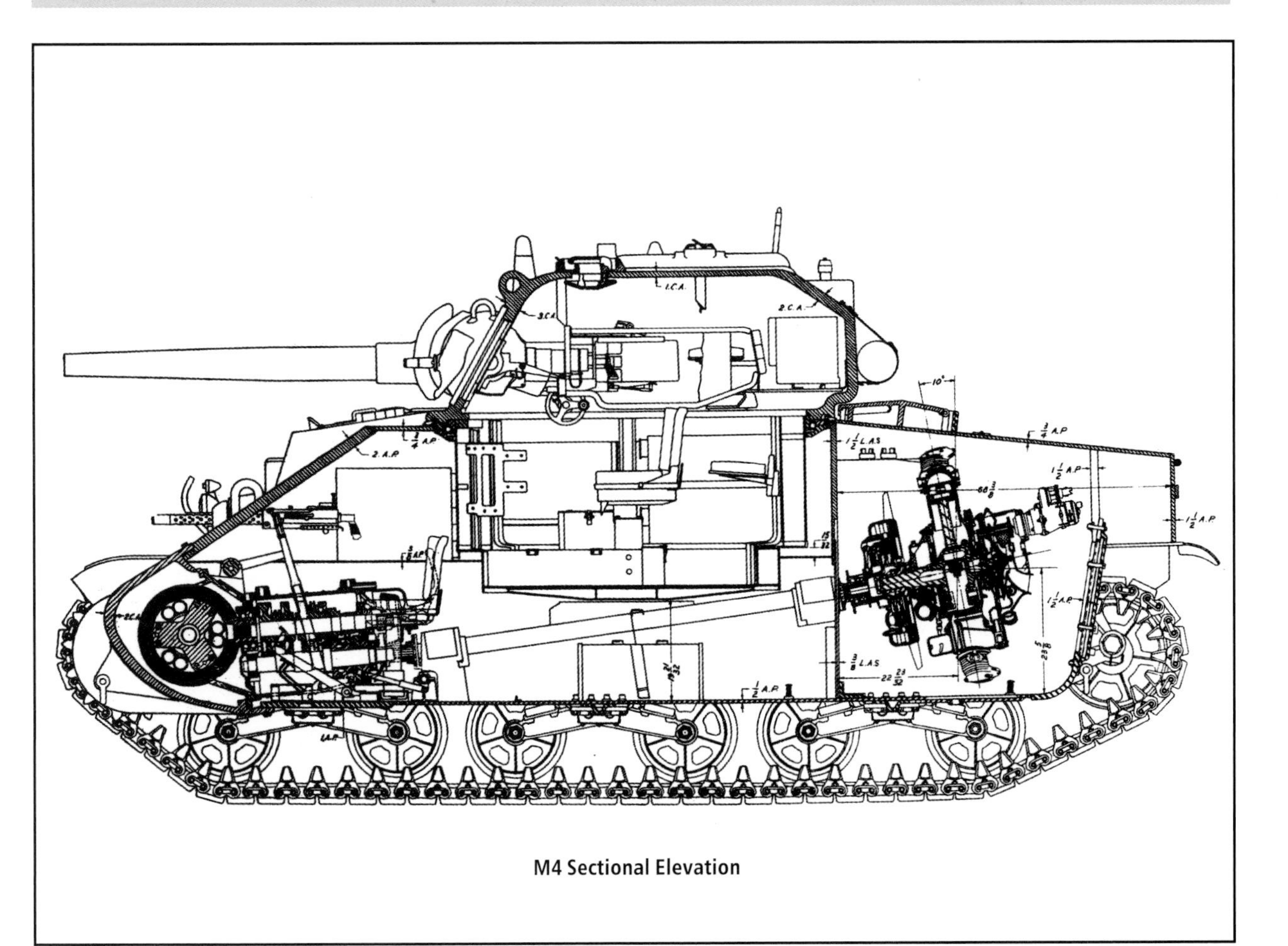

M4 Sectional Elevation

Above:
A British Army Sherman III (M4A2) during the final stages of crossing the River Volturno, Italy, 1943. Note the deep wading trunking on the hull rear. *(TM)*

Right:
The driver of this Sherman III (M4A2) is undergoing training at the RAC Tank Training School at Abbassia, Egypt, during 1943. *(TM)*

Above:
A mid-production M4A3 (British Sherman IV) showing the cast one-piece nose, appliqué armour on the hull sides, and a travelling clamp for the 75mm gun. *(TM)*

Left:
A Sherman OP (observation post) tank with extra aerials and a cable reel on the front hull. Some headquarters or command Shermans were similarly equipped. The location is Italy, July 1944. *(TM)*

CHAPTER FIVE

ARMAMENT

When the Medium Tank M4 series entered production its main 75mm gun was considered adequate for the foreseeable future. That outlook soon changed to the extent that although the 75mm gun was still much in evidence in 1945, new and more powerful 76mm (and other) guns had to be introduced.

One decision that was to blight the early models in the M4 Medium Tank series was made during the late 1930s. That decision was that a 75mm tank gun would be all that was necessary for the foreseeable future. This may have applied during 1939, 1940 and 1941 but by the end of that period it became painfully apparent that no one had informed the Germans of the veracity of this decision. They were busy developing super-powerful 75mm tank and anti-tank guns with a far higher performance than the existing US 75mm M3 tank gun, itself an advanced derivative of a field artillery piece dating back to 1897. In addition, the German's 8.8cm anti-aircraft gun had already demonstrated its anti-armour potency and it was only a matter of time before a dedicated tank gun version appeared, as it duly did on the Tiger.

76mm Gun M1

By late 1942 the need for a more powerful US tank gun had emerged as a high development priority. The immediate proposal was to replace the 75mm M3 gun with a 3in M7 gun, as already installed in the Gun Motor Carriage M10 (Wolverine) tank destroyer. This gun had anti-aircraft origins but was deemed too large and heavy to fit inside the confines of a M4 tank turret (the M10 had an open-topped turret and a voluminous turret bustle). Some trials were carried out with a 3in gun and a M4 turret but they only confirmed what the Rock Island Arsenal designers had forecast.

The decision was therefore to develop a new, lighter and long-barrelled 52-calibre gun capable of delivering more muzzle energy than the 3in M7 gun but firing the same projectiles. To gain the necessary performance the existing 3in projectiles were coupled with a new, longer brass or steel propellant case. For differentiation between the two types of ammunition the new family was designated as 76mm (actually 76.2mm - 3in). The new lightweight gun was standardised during mid-1943 as the 76mm Tank Gun M1 (originally the 76mm Tank Gun T1).

Initial firing trials with the T1 gun were carried out using what was unofficially known as the M4E1 trails vehicle. At first this had a standard turret, later modified to have an enlarged turret bustle to provide more internal space. As a result of these trials a decision was made to

Above:
The added armour on the front of this M4 in Cologne, Germany, March 1944 denotes that it is an M4A3E2 Jumbo assault tank. But unlike most other examples in this series it is armed with a 76mm gun, no doubt taken from a damaged tank. *(TM)*

accommodate the new gun in the larger turret originally designed for the T20/T23 medium tank series that, in time, were meant to replace the M4 series (but never did). The 76mm mantlet and combination mounting were modified to be compatible with existing M4 gun mountings.

Production of the 76mm Tank Gun M1 began during late 1943, with just 200 manufactured by the end of that year. During the whole of 1944 the gun production total reached 8,502, plus another 3,743 during 1945. Not all of these were destined for the M4 (76mm) family for the same gun also formed the main armament of the Gun Motor Carriage M18, Hellcat tank destroyer. When production of the Hellcat ceased in late 1944 the production total had reached 2,507.

This division of the 76mm M1 series gun between two vehicle models meant that after the new gun was introduced to the production lines during February 1944 there were never enough to meet demand. As a result it was not possible to completely replace all the in-service 75mm Gun M4 tanks, as had been hoped at one time. The new gun proved difficult to produce in the numbers required, so despite the apparently prodigious totals mentioned above there was a constant shortfall in supplying and meeting 76mm gun demands throughout 1944. The result was that 75mm gun M4 tanks were still in front-line use when the war ended.

There were four models of 76mm Tank Gun M1, although only three reached the field. The original M1 gun had a plain, unadorned muzzle; only a few were manufactured in this form. With the M1A1 a small counterweight was added to the muzzle to assist the balance of the gyrostabiliser system and to reduce some of the loads on the barrel training mechanisms. In time, most of the M1A1 guns were modified to M1A1C standard. With the M1A1C, stresses imposed on the mounting when firing were alleviated by the addition of a muzzle brake. If the muzzle brake was not installed the screw-on threads around the muzzle were protected by a steel collar. The final form was the 76mm Tank Gun M1A2, also with a muzzle brake but with the right-hand pitch rifling altered from the usual one turn in 40 calibres to one turn in 32 calibres. In theory, during travelling the long barrel was held in a clamp that could fold down onto the hull front when not in use.

The gun breech was of the sliding block pattern with a semi-automatic action in that the

Above: December 1944. Two US Army M4 tanks, still armed with the 75mm gun, prepare to advance towards Bastogne. *(TM)*

breech opened automatically during recoil to eject the spent case. A shield protected the loader during the recoil sequence. Trials demonstrated that a well-trained crew could deliver a fire rate of 20 rounds per minute (rpm), although this was rarely achieved during combat.

The barrel was mounted in a Mount Combination Gun M62, the combination denoting that a 0.30in calibre M1919A4 machine gun was co-axial with the main 76mm gun. Turret movements could be powered electro-hydraulically or under manual control. A full turret traverse of 360° being available. Barrel elevation limits were from -10° to +25°. A gyrostabiliser system kept the gun steady in elevation as the vehicle travelled over rough terrain.

For aiming, the gunner was provided with a Mount, Telescope T116, so arranged that it employed a linkage to maintain the line of vision through a Periscope M10 in the same elevation plane as the gun barrel. Integral with this mounting was a Mount, Telescope M47 carrying a Telescope M71 or M83. Using this arrangement the telescope could be moved independently of the barrel and periscope for gun laying. The telescope sight graticules were calibrated for up to 3,000yd (2,743m). Firing was electrical via a solenoid operated by one of two foot pedals to the left of the driver, the second foot pedal being used to fire the co-axial machine gun.

The manufacture and supply of 76mm ammunition also gave rise to shortfalls in deliveries throughout 1944. One of the reasons for the ammunition difficulties was that it had been decided to introduce a new type of armour-piercing projectile known as High Velocity Armour Piercing, Tracer (HVAP-T). With this fixed round (ie the projectile was crimped into the case around the projectile drive band) the main armour-piercing element was a sub-calibre penetrator formed from a hardened, dense

tungsten carbide penetrator core held within an aluminium sheath. A slightly heavier propellant charge than normal imparted a high muzzle velocity, enhancing anti-armour penetration performance. The nominal muzzle velocity with HVAP-T was of the order of 3,399ft/s (1,036mtr/s). However, this round was found to be difficult to manufacture in the quantities required so reliance continued to be made on armour-piercing capped projectiles (APC M62) with a muzzle velocity of 2,598.4ft/s (792mtr/s). Using this latter round it was possible to penetrate 3.94in (100mm) of armour at 1,006yd (915mtr), rendering the carrier tank capable of engaging German Panther and Tiger tanks with a reasonable expectancy of success.

Also available were impact-fused, high-explosive (HE M42A1) and base ignition smoke (Smoke M8) projectiles, with an illuminating projectile appearing after World War Two. All these rounds were of the fixed type.

Ammunition stowage within the vehicle was another protection area that had to be altered following combat experience. The first members of the Medium Tank M4 series carried their ammunition in racks along the sides of the combat compartment, with more rounds scattered around the interior and the floor. Due to the relatively thin side armour these racks were vulnerable to incoming fire and were one of the main reasons why the M4 family was so prone to 'brewing up' (catching fire) when hit. By the time the M4 (76mm) tanks reached the production stage in February 1944, provision had been provided for the rounds to be arranged within 'wet stowage'. The term indicated that the rounds were protected behind a wall barrier containing a mixture of glycerine and water. This stowage measure did not entirely do away with ammunition fires but did reduce them significantly. The usual 76mm ammunition combat load was 71 rounds.

76 mm Gun M1A1C data

Calibre:	76.2 mm
Overall length:	13.65ft (4.16m)
Length of barrel:	13ft (3.962m)
Length of rifling:	11.13ft (3.392m)
Length of projectile travel in bore:	11.28ft (3.437m)
Length of recoil:	12.51in to 14.4in (317.5 to 356mm*)
Rifling:	right-hand twist, one turn in 40 calibres
Weight of barrel and breech mechanism:	1,205.92lb (547kg)

*dependent on ammunition type

76mm Ammunition Data

	HVAP-T	APC	HE	Smoke
Round weight:	18.9lb (8.577kg)	22.13lb (10.04kg)	22.57lb (10.24kg)	22.66lb (10.2785kg)
Projectile weight:	9.4lb (4.26kg)	15.43lb (7kg)	12.57lb (5.84kg)	12.94lb (5.87kg)
Propellant charge weight:	3.9lb (1.77kg)	3.75lb (1.77kg)	3.75lb (1.77kg)	3.75lb (1.77kg)
Muzzle velocity:	3,398.92ft/s (1,036m/s)	2,598.4ft/s (792m/s)	2,700.1ft/s (823m/s)	2,700.1ft/s (823m/s)

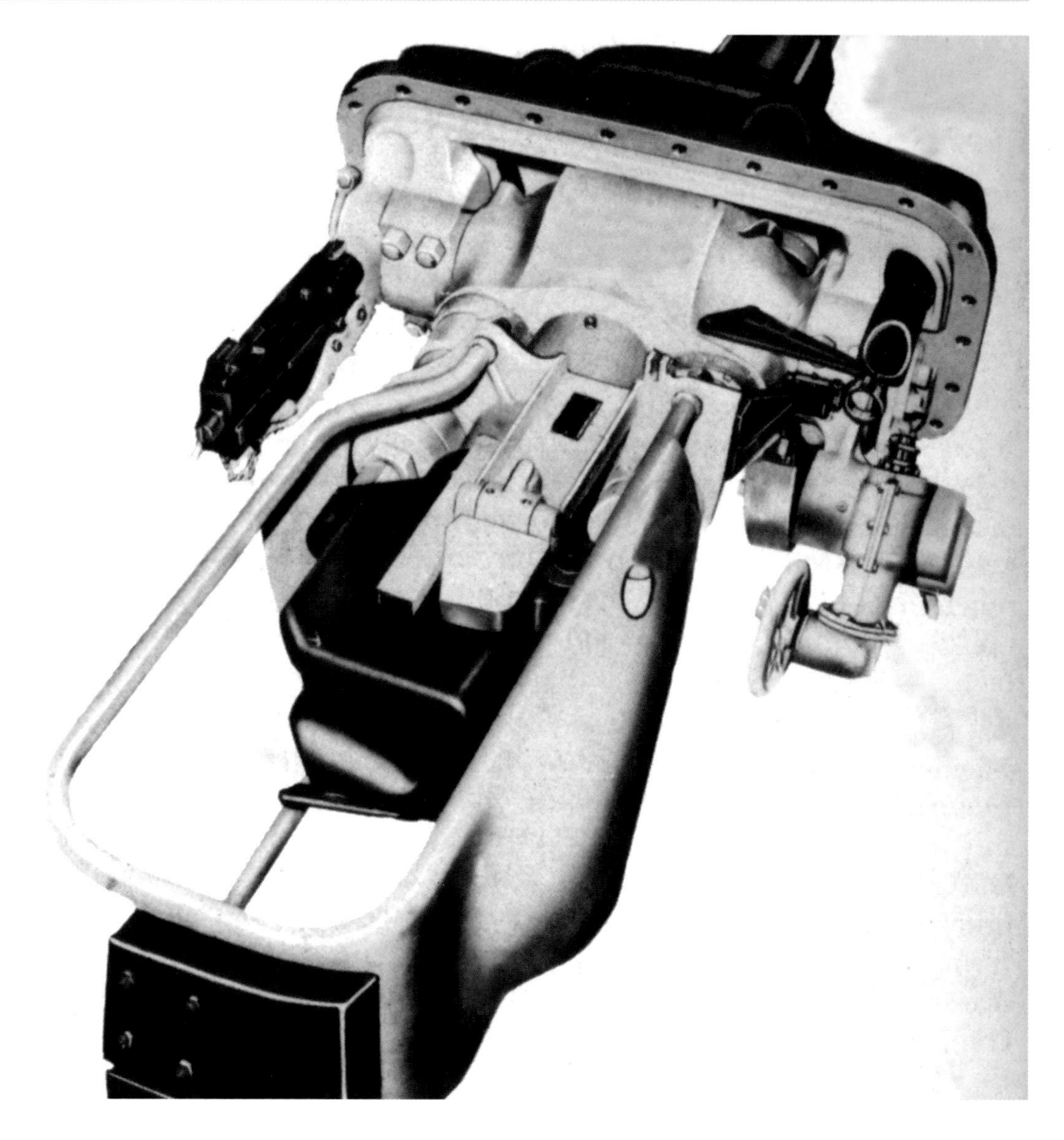

Right:
The breech of the 75mm Gun in a Gun Mounting M34. To the left of the breech is the .30-calibre co-axial machine gun. *(TM)*

Above: The gun mantlet on this early production M4 series vehicle has its 75mm M3 gun in a Gun Mount M34 with a small shield covering the mantlet. On the later Gun Mount M34A1 this shield was enlarged. *(TM)*

105mm Howitzer M4

By 1942 it had been accepted that armoured formations had to have their own integral artillery support vehicles at Medium Tank Battalion Headquarters level to keep pace with the tempo of operations. One such vehicle emerged as the Howitzer Motor Carriage M8, an M5 light tank chassis carrying a stubby 75mm howitzer in an open turret. Despite the success of this conversion the shell power and limited range of the howitzer was considered insufficient for the armoured fire support role so consideration was given to mounting a 105mm howitzer in a Medium Tank M4 series turret.

This consideration commenced in November 1942 with trials carried out using the M4A4E1 and M4E5 test vehicles. After modifications had been introduced, the result was standardised for production in August 1943 as the M4 (105mm) and M4A3 (105mm). Production did not start until early 1944.

The howitzer selected, the 105mm Howitzer M4, was a modification of the towed 105mm Howitzer M2A1, still in widespread service all around the world in the 21st century, although now known as the M101. By turning the breech block on its side to slide open horizontally to the right, instead of vertically, and by repositioning the breech operating lever it was possible to mount this 22.5-calibre howitzer within the confines of a normal M4 series turret. In addition, firing became electrical via a solenoid, although a lanyard-actuated firing mechanism was provided as a back-up. Four rounds could be fired within 30 seconds, although for more prolonged firings the firing rate was reduced to 30 rounds in 10 minutes.

The full title for the howitzer mounting was the Howitzer, 105mm, M4; Mount, Combination Gun M52. Both M4 and M4A3 models were converted for the howitzer role at the Detroit Tank Arsenal. A total of 2,286 vehicles were produced during 1944 and a further 2,394 during 1945, bringing the final total to 4,680. Once sufficient M4 (105mm) vehicles were available the smaller Howitzer Motor Carriage M8 was phased out of service.

Production figures for the Howitzer M4 were 168 during 1943, 2,404 during 1944 and 2,563 during 1945, a total of 5,135. This was in excess of the M4 (105mm) tanks produced (4,680) so the surplus was held as spares.

The turret, combination mount and firing arrangements remained much the same as for the M4 (75mm) vehicles. One of the main changes was that the M4 (105mm) did not have a powered turret traverse while another was the absence of a gyrostabiliser as most firings were carried out from the static position. In addition the barrel elevation limits were from -10° to +35°. A full 360° turret traverse was maintained.

The 105mm ammunition fired from the Howitzer M4 remained as for the towed M2A1 howitzer. Unlike 76mm gun ammunition, it was described as semi-fixed as the propellant charge for the howitzer could be altered to suit any specific fire mission. Up to seven bagged propellant charges could be selected before the projectile and brass or steel propellant cases were

Sherman M4A3 (76mm)

Scale 1:35

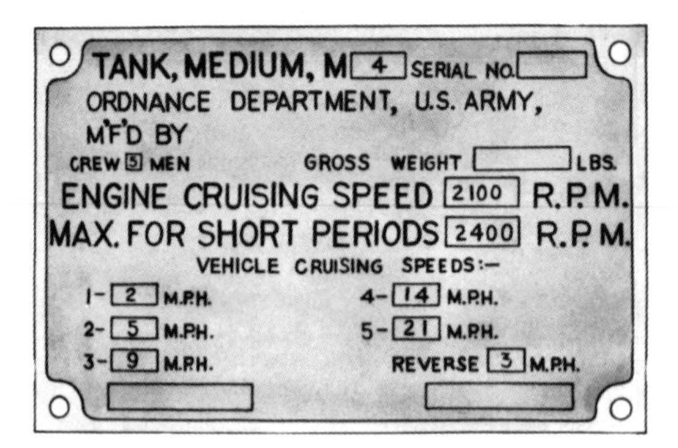
TANK, MEDIUM, M 4 SERIAL NO.
ORDNANCE DEPARTMENT, U.S. ARMY,
M'F'D BY
CREW 5 MEN GROSS WEIGHT LBS.
ENGINE CRUISING SPEED 2100 R.P.M.
MAX. FOR SHORT PERIODS 2400 R.P.M.
VEHICLE CRUISING SPEEDS:–
1- 2 M.P.H. 4- 14 M.P.H.
2- 5 M.P.H. 5- 21 M.P.H.
3- 9 M.P.H. REVERSE 3 M.P.H.

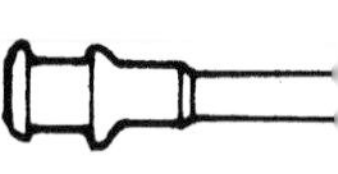

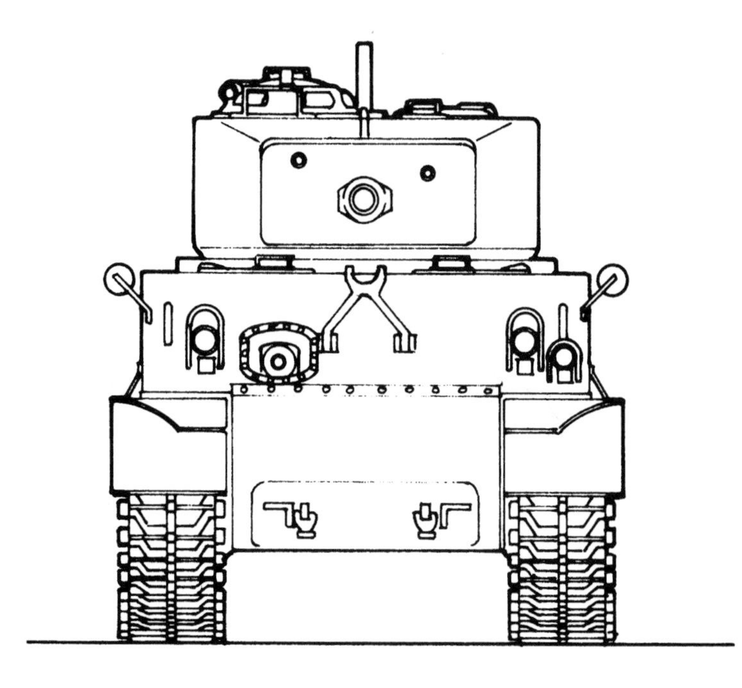

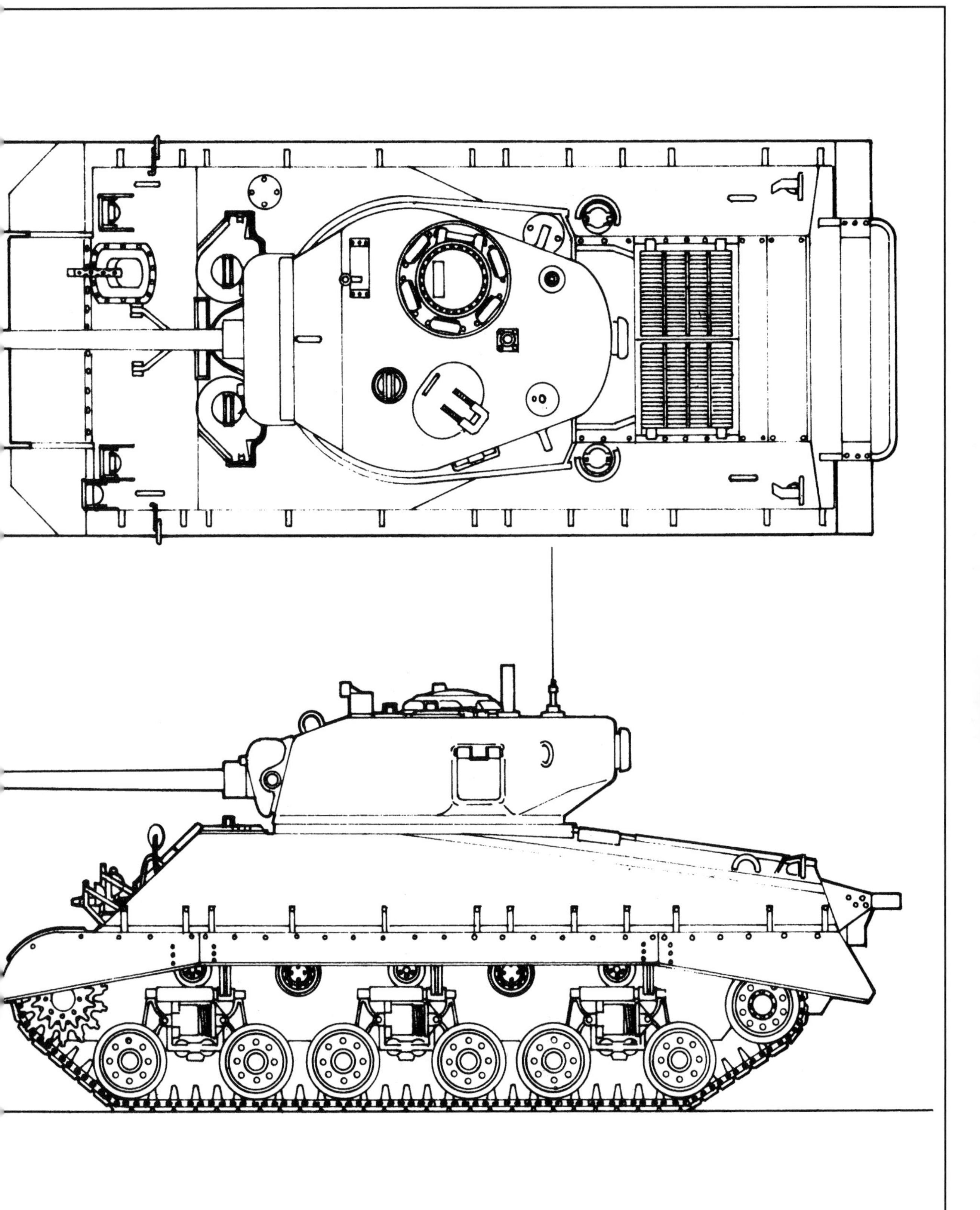

Above:
This view of a training rig for an M4 series tank is taken from under the gun breech area and looking towards the rear. *(TM)*

Right:
Looking down through the turret ring of an M4 series tank with the lower floor removed to expose the stowage positions for machine gun and sub-machine gun ammunition. *(TM)*

Above:
A training rig for an M4 series tank indicating some of the main ammunition and other stowage points. *(TM)*

mated on a loading frame just prior to loading. There was internal stowage for up to 66 rounds of 105mm ammunition. A further 44 rounds could be carried in an Armored Trailer M8 towed from a rear-mounted hitch on the 105mm tanks, but it appears that this facility was little used.

As the M4 (105mm) was primarily a close fire support platform, two main ammunition types were fired, high explosive (HE M1) and Smoke, both projectiles weighing around 33.07lb (15kg). The smoke could be white for smoke screens or yellow, green or red for target marking purposes. Also available were chemical rounds, although they were never used in combat. Using the maximum Charge 7, producing a muzzle velocity of 1,548.54ft/s (472m/s), the maximum possible range was about 12,205yd (11,160m). Most fire missions carried out by the M4 (105mm) tanks involved shorter ranges.

Machine guns and ancillaries

Both the M4 (76mm) and M4 (105mm) tanks carried the same machine gun armament, namely two 0.30in Machine Gun M1919A4s and a single 0.50in Machine Gun M2 HB (HB - heavy barrel). Both types of weapon had air-cooled barrels. One of the Browning 0.30in M1919A4s was co-axial with the main armament and fired electrically via a solenoid, although the manual trigger mechanism was retained. The second was a bow machine gun in a flexible ball mounting operated by the co-driver and fired using a trigger and pistol grip. On the 105mm tanks ammunition stowage was provided for 4,000 rounds of .30in ammunition in belts (6,250 rounds on M4 (76mm)

105 mm Howitzer M4 data

Calibre: 105mm
Overall length: 8.45ft (2.575m)
Length of bore: 7.75ft (2.363m)
Length of projectile travel in bore: 6.81ft (2.075m)
Length of recoil: up to 13.99in (355mm)*
Rifling: right-hand twist, one turn in 20 calibres
Weight of barrel and breech mechanism: 1,139.78lb (517kg)

*dependent on ammunition type

Right:
Head-on view of the turret for an M4A1E8 (76mm) turret, clearly showing the location of the roof hatches for this model. Also to be seen are the drivers' hatches. *(TM)*

Left:
Top view of a late production M4 series commander's cupola showing the all-round vision provided by the use of six episcopes. *(TM)*

Left:
On some Sherman series tanks the British Army chose to install its own type of commander's cupola hatch with a two-piece lid construction. *(TM)*

Right:
A Gun Motor Carriage M10, based on the hull of the M4 series but with an open-topped turret, armed with a 3-inch gun. As well as all the extra protection carried, the vehicle is equipped with a Rhinoceros device. *(TM)*

tanks). Also carried on the vehicle was a Tripod Mount M2 onto which one of the .30in machine guns could be placed for dismounted action.

The 0.50in Machine Gun M2 HB was also a Browning product. Intended for air defence this gun could also be used against ground targets. On early models this machine gun was fitted on a pintle mount on the commander's cupola, although the pintle was later transferred to the turret roof alongside the cupola hatch. For the M4 (105mm) tank ammunition stowage was provided for three hundred 0.50in rounds in belts (600 rounds on the M4 [76mm]).

The turret armament was completed by a 2in Mortar M3 in the roof and facing forward. The mortar was fixed and manually fired to create a rapid smoke screen when necessary. Eighteen Smoke Mk 1 grenades were carried for this mortar.

Right:
Apart from Medium Tank M4 series vehicles armed with the M1 series 76mm gun, the other main carrier was the Gun Motor Carriage M18 Hellcat, a potent and fast tank destroyer. *(TM)*

Left:
A Medium Tank M4A3 (76mm) operating in France, August 1944, armed with an M1A1 76mm gun and well laden with extra protection against German 8.8cm *Raketenpanzerbüchse* shaped-charge warhead rockets. *(TM)*

On paper, the M4 (105mm) crew's personal defensive armament was limited to a single 0.45in sub-machine gun, usually an M3 or M3A1 'Grease Gun', although this official total was often exceeded. On the M4 (76mm) stowage was provided for five of these weapons. Many crew members also carried a sidearm, typically a 0.45in M1911A1 automatic pistol. Ammunition stowage was provided for 600 rounds of 0.45in ammunition (900 rounds on the M4 [76mm]).

In addition to all this weaponry each tank had stowage for 12 hand grenades, six high-explosive fragmentation (HE-FRAG II) and six Smoke M15. Two thermite grenades could be carried to wreck the gun chamber and barrel in extreme cases where capture of the vehicle by an enemy was inevitable.

Left:
Heaviest of all Shermans was the 'Jumbo'. Known as the M4A3E2, over 250 were built by Grand Blanc between May and June 1944. These were rushed to Europe where they performed very well. A few were fitted with 76mm guns but most had the standard 75mm. *(TM)*

Right:
Aboard an LCT, prior to the Normandy landings. Just visible in the foreground is an M4A1. The tank with all the travelling and deep wading equipment is an M4A2. *(SZ)*

U.S.A.-3036947

CHAPTER SIX

FIREFLY

One of the most remarkable improvisations of the M4 tank series was the Sherman Firefly, a British measure in which a 17-pounder anti-tank gun was mounted into the tight confines of an M4 turret normally mounting the 75mm gun.

The British were as tardy as the Americans in recognising the need for increased calibres for tank armament prior to 1942, considering for too long that a 75mm gun was all that would be needed. Fortunately the Royal Artillery, then responsible for anti-tank gun activities, realised that by 1941 its planned 6-pounder anti-tank guns would rapidly become obsolete thanks to increases in the thickness of German tank armour. They therefore planned to make a significant move in development terms by adopting a 3in (76.2mm) 58-calibre anti-tank gun that would, it was hoped, remain effective for years. Selection of the 3in calibre was prompted by the existence of manufacturing tooling and machinery that had previously been used to manufacture anti-aircraft and naval guns. Known as the 17-pounder, the first towed examples were delivered as early as August 1942.

The 17-pounder also came under scrutiny as a potential tank gun. As all existing British tank designs then had hulls that were too narrow to accommodate the large turret ring needed for the 17-pounder, the original intention was to mount the gun in a purpose-built tank based around Comet tank components and known as the Cruiser Tank, Challenger (A30). As a safeguard it was also decided, around January 1943, to investigate the possibilities of installing the 17-pounder in a Lend-Lease Sherman tank. It was just as well that this measure was taken for early trials with the Challenger prototypes indicated that it was generally unsatisfactory, so the Sherman project was awarded a higher priority from the last quarter of 1943 onwards, especially as the invasion of Europe was scheduled for mid-1944. A demonstrator model was ready by November 1943. By February 1944 it became painfully apparent that the Challenger project would not reach fruition in time for the planned operations in Europe so the 17-pounder/ Sherman combination was lifted to a higher priority rating.

The result was the most powerful British tank design of the war years, the Sherman Firefly, usually known simply as the Firefly. Much of the original Medium Tank M4 was left unaltered for the conversion; the main centre of engineering related to the turret. Although a 17-pounder gun could be shoehorned into the existing turret, the size of the breech mechanism and the length of recoil meant there was no space left for the usual radio equipment installation at the rear of the turret. This drawback was overcome by simply cutting a

Above:
Tanks of the Scots Greys occupying the centre of Wismar in early 1945. Amid the debris of war, the front vehicle is a Firefly (Sherman IIC). *(TM)*

Left:
Reloading a Sherman VC (Firefly) with 17-pounder armour-piercing ammunition, clearly indicating the bulk of the one-piece rounds. Note the German helmet trophy on the front hull. *(TM)*

British, National Identification Marks

Up to 1942 (Western Desert).

1942-45.

British and Commonwealth formation signs

1st Armoured Division.

2nd Armoured Division.

6th Armoured Division.

7th Armoured Division, 1940-44.

7th Armoured Division, 1944-45.

79th Armoured Division "The Funnies".

8th Armoured Division.

9th Armoured Division.

10th Armoured Division.

British and Commonwealth formation signs

11th Armoured Division.

Guards Armoured Division.

6th (Guards) Tank Brigade.

4th Canadian Armoured Division.

5th Canadian Armoured Division.

1st Canadian Armoured Brigade.

Australian Expeditionary Force, Middle East.

Australian 1st Armoured Division.

Australian 2nd Armoured Division.

Australian 3rd Armoured Division.

New Zealand 4th Armoured Brigade.

Indian, 31st Armoured Division.

Right:
The overall layout of a Sherman VC Firefly showing the length of the gun barrel and the absence of the hull machine gun position. This is a late war trials tank with hull-side fittings that were not carried on active service. *(TM)*

rectangular hole in the rear of the turret and welding an armoured box over the aperture. The armoured box then acted as the radio housing and also as a counterweight to the increased forward weight of the gun installation. An access and escape hatch for the loader was fitted into the turret roof as the bulk of the gun and breech precluded the usual escape route through the commander's hatch. This feature was later adopted for late production M4s.

Other internal modifications related mainly to ammunition stowage. The 17-pounder fixed rounds were longer than those for the US Army's 76mm Gun M1 yet the number that could be carried internally was 78 rounds. To provide space for that many rounds meant removing the co-driver position and bow machine gun, limiting a Firefly crew to four. The resultant space was then used for ammunition stowage. An armoured plate was welded over the bow machine-gun ball mounting in the hull front. If the co-driver station had been retained the number of 17-pounder rounds carried would have been too limited for tactical comfort. Soon after the war ended most

Below:
Overhead view of the pilot model for the Firefly. The armoured box added to the rear of the turret can be seen, as can a great deal of turret detail. *(TM)*

Above:
Starting the last year of the war with anticipation, a Firefly on the Franco-German border on 1 January 1945 is so well dug in that only the turret shows. *(TM)*

new tank designs omitted the bow machine-gun position altogether, considering that it was too limited in combat effectiveness and required an extra crew member at a time when the trend was to limit the size of tank crews wherever possible. The Firefly was one of the first battle tank designs not to have a bow machine gun. One further Firefly modification was a stowage box on the rear of the hull.

About 600 Shermans were converted to Firefly standard. Most of the vehicles involved were Sherman Vs, M4A4s with the Chrysler Multibank engine. They then became the Sherman VC (Firefly). Also involved in the Firefly programme was the Sherman IIC (Firefly), based on the M4A1. Fireflys were always provided with extra appliqué armour plates welded on the hull sides.

The 17-pounder gun installed in the Firefly was the Ordnance QF (quick-firing) 17-pounder Mk IV or Mk VII, the only difference between the two Marks the form of the recoil mechanism casing. A muzzle brake was fitted, as on the towed gun, and to ensure the gun would fit into the Sherman turret it was turned on its side so that the sliding breech block operated horizontally. In addition the length of recoil was limited to just 15.01in (381mm) instead of the more usual 3.31ft (1.01m). For travelling, the barrel was initially traversed over the rear of the vehicle and held in a folding clamp to prevent the long barrel striking structures or trees. After combat experience the clamp was moved to the hull front, as on the M4 (76mm) tanks.

Ammunition carried by the Firefly was usually limited to armour-piercing ammunition as the vehicle's primary role was the destruction of enemy armour. Impact-fused high-explosive rounds could become involved for direct fire support missions. The armour-piercing rounds involved either solid steel armour-piercing with a

17-pounder Gun data

Calibre:	76.2 mm
Overall length:	14.58ft (4.443m)
Length of bore:	13.78ft (4.2m)
Length of rifling:	11.68ft (3.562m)
Length of recoil:	15.01in (381mm)
Rifling:	right-hand twist, one turn in 30 calibres
Weight of barrel and breech mechanism:	4,479.75lb (2,032kg)

United States, National identification marks.

Up to late 1941.

1941-43.

Tank turret usage.

Allied air recognition signs.

Allied air recognition.

Allied air recognition, variant.

United States tactical marks.
13th Armored Regiment, 1st Armored Division, Tunisian campaign.

US 1st Armored Division.

Combat Command HQ. (outline triangles occasionally appear solid.)

M4A2 (76mm) of the 7th US Armored Division in winter camouflage, Colmar, France, 1944. *(TM)*

Free Forces

French

National Identification Mark, North Africa.

National Identification Mark, from 1943.

National Identification Mark variant, from late 1943.

Free French Forces.

2nd Armoured Division, 1945.

Polish

1st Polish Corps.

1st Armoured Division.

2nd Armoured Division.

Free Forces Formation Signs

1st Independent Belgian Brigade Group.

Czech Independent Armoured Brigade Group.

Jewish Brigade Group.

Above:
The pilot model of the Sherman IIC, the Firefly. The armoured box turret extension to the rear is very apparent. *(TM)*

tracer element (AP-T) or armour-piercing capped, also with a tracer (APC-T). As the gun designation implied, the projectile weighed approximately 17lb (7.7kg) or slightly less. Several marks of each type of projectile were available, most of them differing only in detail or source of manufacture.

Projectiles were fired at a nominal muzzle velocity of 3,116.8ft/s (950m/s). The maximum possible range was about 9,999.88yd (9,144m) although the maximum effective anti-tank range was about 2,001.3yd (1,830m). Even at that range an APC-T shot could penetrate 3.58in (91mm) of armour. Also using APC-T it was possible to defeat 5.12in (130mm) of armour set at an angle of 30° at 1,000yd (914m). This performance enabled the 17-pounder Firefly to tackle heavy German tanks such as the Panther or Tiger on something approaching equality in gunnery terms (although not in protection terms). It was the only British tank able to do so, right until the last days of the

Right:
The Achilles IIC, the British version of the Gun Motor Carriage M10 armed with a British 17-pounder in place of the US 3in gun. *(TM)*

Left:
Intended to be the main British 17-pounder gun tank for the invasion of Europe in 1944, the A30 Challenger proved to be difficult to get into production in time for the attack, leading to the development of the Firefly. *(TM)*

war yet its armoured protection remained as inadequate as for the other M4/Sherman models.

The Firefly also suffered in being available only in limited numbers. By early 1944 the towed 17-pounder was much in demand for anti-tank batteries to defend field formations and also for tank destroyers in British service. These included the Wolverine (the British conversion of the US Gun Motor Carriage M10 to mount the 17-pounder) and the Valentine Archer. The supply of guns for the Firefly conversions became so limited that in the immediate aftermath of the Normandy landings the allocation of the Firefly was limited to one tank in every troop of four tanks. When sufficient vehicles later became available this scale of issue was increased.

The Firefly may have been an expedient design but it proved to be highly successful.

Left:
Another British improvisation to get a 17-pounder tank into service was the Valentine Archer. For this tank destroyer vehicle the gun pointed to the rear and was carried in a limited traverse mounting. It proved to be remarkably successful. *(TM)*

Above:
A well preserved example of a Sherman VC Firefly. *(JB)*

Right:
This example of a Sherman VC Firefly has the late production hull front but retains the three-piece nose. Note the absence of the hull machine gun. *(JB)*

Above:
Now a collector's piece, this late production M4A1E8 (76mm) is armed with a 76mm Gun M1. *(JB)*

Left:
Carefully preserved in every detail, a late production M4A1E8 (76mm), complete with HVSS. *(JB)*

Above: Rumbling on to a new service career in the post-war world, a Medium Tank M4 in Southern Germany, March 1944. *(TM)*